Study Guide

Nutrition
Concepts and Controversies

ELEVENTH EDITION

Frances Sizer

Ellie Whitney

Prepared by

Jana R. Kicklighter
Georgia State University

 **WADSWORTH**
CENGAGE Learning

Australia • Brazil • Japan • Korea • Mexico • Singapore • Spain • United Kingdom • United States

WADSWORTH
CENGAGE Learning

**Study Guide for Nutrition Concepts and Controversies,
Eleventh Edition**
Frances Sizer, Ellie Whitney

For product information and technology assistance, contact us at
**Cengage Learning Customer & Sales Support,
1-800-354-9706**

For permission to use material from this text or product, submit all
requests online at **www.cengage.com/permissions**
Further permissions questions can be emailed to
permissionrequest@cengage.com

ISBN-13: 978-0-495-55304-5
ISBN-10: 0-495-55304-2

Wadsworth
10 Davis Drive
Belmont, CA 94002-3098
USA

Cengage Learning is a leading provider of customized learning
solutions with office locations around the globe, including Singapore, the
United Kingdom, Australia, Mexico, Brazil, and Japan. Locate your
local office at: **international.cengage.com/region**

Cengage Learning products are represented in Canada by
Nelson Education, Ltd.

For your course and learning solutions, visit
academic.cengage.com

Purchase any of our products at your local college store or
at our preferred online store **www.ichapters.com**

Printed in the United States of America
2 3 4 5 6 7 11 10 09 08

Table of Contents

Preface

This Study Guide has been designed to accompany the eleventh edition of *Nutrition: Concepts and Controversies* and its chapter divisions parallel those in the text. The primary goal of the Study Guide is to facilitate your study of the material presented in the text and enhance your learning. Pages in the Study Guide have been perforated and three-hole punched to enable you to insert, delete, or rearrange pages to best suit your needs or to correspond with the sequence used by your instructor.

Each chapter consists of seven sections: chapter objectives, key concepts, a fill-in-the-blank chapter summary, chapter glossary of key terms, exercises which include chapter study questions, short answer questions, problem-solving/application questions and controversy questions, sample test items, and an answer key. In addition, some chapters have study aids that complement material presented in the text.

To obtain the maximum benefit from this Study Guide you should: (1) read the chapter in the text; (2) study your class notes, and (3) complete the corresponding chapter in the Study Guide, including the following sections:

- **Chapter Objectives and Key Concepts.** Read the objectives and key concepts at the beginning of each chapter to help you focus your attention on the most essential material and concepts presented in the chapter.

- **Summing Up.** This section provides a brief summary of the chapter using a fill-in-the-blank format. To help yourself summarize and review the main points in the chapter, read the summary and fill in the blanks.

- **Chapter Glossary.** This section includes key terms and their definitions presented in the margins of the text. Match the definitions and terms or complete the crossword puzzles to enhance your understanding of the material and prepare for many of the test items.

- **Exercises.** Chapter study questions, short answer questions, problem-solving/application questions and controversy questions are included in this section. Chapter study questions and controversy questions allow you to review key concepts and help you prepare for essay questions on examinations, while short answer and problem-solving/application questions help you to review and apply key concepts and principles presented.

- **Study Aids.** Many chapters include study aids to help you remember, understand and practice key information presented in the text. Some study aids refer to tables and figures in your textbook.

- **Sample Test Items.** Comprehension-level and application-level multiple choice, matching, and true/false items are included to help you test your recall and understanding of the basic content presented in the text. Many of the questions are likely to be similar to those included on examinations. However, rather than simply learning the answers to the sample test questions, you should use the questions in this section as the basis for learning and studying important concepts and principles presented in the chapter.

- **Answers.** Answers are provided for all sections of the Study Guide. When checking fill-in-the-blank answers, be aware that more than one word may satisfactorily complete a sentence. If your answer differs from the one provided, determine if you have used an acceptable synonym or whether you missed the basic concept. When checking chapter study questions and controversy questions, keep in mind that only key points have been included in the answer. Good answers to discussion questions require not only a recall of the key points but also descriptions of interrelationships and explanations of why and how. A well-written answer to discussion questions requires synthesis of knowledge gained from previous lessons, related courses, and personal experience. After completing the sections of the chapter, check your answers and use your results to reinforce and strengthen the learning process by going back and restudying areas that you were unsure of or missed.

We hope you find this Study Guide to be useful. Please let us hear from you if you have suggestions for ways to improve the Study Guide.

Jana Kicklighter, Ph.D., R.D.
Division of Nutrition
Georgia State University
Atlanta, Georgia 30303

associations, emotional comfort, values or beliefs, region of the country, weight and

(18)_____ value. A key to wise diet planning is to make sure that the foods you eat daily,

your (19)_____ foods, are especially nutritious.

The U.S. Department of Health and Human Services sets ten-year health objectives for the nation in

its document (20)_____. In addition to nutrition, (21)_____ also plays

a prominent role in this document.

An easy way to obtain a nutritious diet is to consume a variety of fruits, vegetables, meats,

(22)_____ and milk products each day. The extent to which foods support good health

depends on the (23)_____, nutrients and nonnutrients they contain. A nutritious diet has

five characteristics, including (24)_____, balance, calorie control,

(25)_____, and variety. To act on nutrition knowledge, people must change their

(26)_____. Obstacles to behavior change, which can cause lapses, include competence,

(27)_____ and motivation.

Chapter Glossary

Crossword Puzzle:

Across:	*Down:*
2. a measure of nutrients provided per calorie of food	1. styles of cooking
4. people who exclude from their diets animal flesh and possibly other animal products such as milk, cheese, and eggs	3. diets composed of purified ingredients of known chemical composition
5. people who eat foods of both plant and animal origin, including animal flesh	6. the dietary characteristic of providing constituents within set limits, not to excess
7. any condition caused by excess or deficient food energy or nutrient intake or by an imbalance of nutrients	10. units of energy
8. long-duration degenerative diseases characterized by deterioration of the body organs	11. the dietary characteristic of providing all of the essential nutrients, fiber, and energy in amounts sufficient to maintain health and body weight
9. the dietary characteristic of providing foods of a number of types in proportion to each other, such that foods rich in some nutrients do not crowd out of the diet foods that are rich in other nutrients	
12. control of energy intake, a feature of a sound diet plan	
13. the dietary characteristic of providing a wide selection of foods—the opposite of monotony	
14. units of weight	

Word Bank:

adequacy	cuisines	nutrient density
balance	elemental diets	omnivores
calorie control	grams	variety
calories	malnutrition	vegetarians
chronic diseases	moderation	

Matching Exercise:

_____ 1. nutrition

_____ 2. essential nutrients

_____ 3. food

_____ 4. nonnutrients

_____ 5. organic

_____ 6. foodways

_____ 7. energy-yielding nutrients

_____ 8. nutrients

_____ 9. energy

_____ 10. legumes

_____ 11. phytochemicals

_____ 12. dietary supplements

_____ 13. ethnic foods

_____ 14. diet

_____ 15. genome

_____ 16. self efficacy

_____ 17. genes

_____ 18. lapses

_____ 19. locus of control

a. components of food that are indispensable to the body's functioning; they provide energy, serve as building material, help maintain or repair body parts, and support growth

b. carbon containing

c. the study of the nutrients in foods and in the body

d. the foods (including beverages) a person eats and drinks

e. nonnutrient compounds in plant-derived foods having biological activity in the body

f. the sum of a culture's habits, customs, beliefs, and preferences concerning food

g. foods associated with particular cultural subgroups within a population

h. the capacity to do work

i. the nutrients the body cannot make for itself (or cannot make fast enough) from other raw materials

j. the nutrients the body can use for energy

k. pills, liquids, or powders that contain purified nutrients or other ingredients

l. medically, any substance that the body can take in and assimilate that will enable it to stay alive and to grow

m. beans, peas and lentils valued as inexpensive sources of protein, vitamins, minerals and fiber that contribute little or no fat to the diet

n. compounds other than the six nutrients that are present in foods and that have biological activity in the body

o. the full complement of genetic material in the chromosomes of a cell

p. times of falling back into former habits

q. units of a cell's inheritance

r. the assigned source of responsibility for one's life's events

s. belief in one's ability to take action and successfully perform a behavior

Exercises

Answer these chapter study questions:

1. What is meant by the statement that protein does double duty?

2. What is meant by the term essential nutrients?

3. What does food provide in addition to nutrients?

4. Why do nutrition scientists agree that people should not eat the same foods day after day?

5. Why does the number of foods available today make it more difficult, rather than easier, to plan nutritious diets?

Complete these short answer questions:

1. The four organic nutrients include:

   a.   c.

   b.   d.

2. The six kinds of nutrients include:

   a.   d.

   b.   e.

   c.   f.

3. The energy-yielding nutrients are:

   a.   c.

   b.

4. Factors people cite to explain food choices include:

   a.   h.

   b.   i.

   c.   j.

   d.   k.

   e.   l.

   f.   m.

   g.   n.

5. Five characteristics of a nutritious diet include:

   a.   d.

   b.   e.

   c.

6. The six stages of behavior change include:

   a.   d.

   b.   e.

   c.   f.

Solve these problems:

_____ 1.	I always drink coffee at morning break.	a. personal preference
_____ 2.	I usually microwave a frozen dinner after getting home from my night class.	b. habit
_____ 3.	All my friends are going out for pizza after the game.	c. social pressure
_____ 4.	I use dried beans as a substitute for meat because meat costs too much.	d. availability
		e. convenience
_____ 5.	I eat a lot of meat because I like the taste.	f. economy
_____ 6.	I usually choose chicken rather than beef as a meat because it is healthier.	g. nutritional value
_____ 7.	I have to eat high-fat foods because I eat in the employee cafeteria every day for lunch.	

Answer these controversy questions:

1. What type of screening is required before research findings are published in scientific journals?

2. What criterion would denote that a physician is well-qualified to speak on nutrition?

3. What is a diploma mill?

4. What are the qualifications of a dietetic technician?

5. Which type of dietetic subspecialties would you expect to see in a hospital setting?

6. What is one of the most trustworthy websites for scientific investigation and what does it provide?

Study Aid

Use Table 1-3 on page 6 in your textbook to familiarize yourself with the elements in the six classes of nutrients.

Sample Test Items

Comprehension Level Items:

1. A scientist's responsibilities include all of the following **except**
   a. obtaining facts by systematically asking questions.
   b. conducting experiments to test for possible answers.
   c. submitting their findings to the news media.
   d. publishing the results of their work in scientific journals.

2. Which of the following nutrients is **not** organic?
   a. minerals
   b. vitamins
   c. protein
   d. carbohydrates
   e. fat

3. Foremost among the six classes of nutrients in foods is:
   a. fats.
   b. carbohydrates.
   c. vitamins.
   d. water.

4. Which of the following is **not** an energy-yielding nutrient?
   a. carbohydrates
   b. vitamins
   c. fats
   d. protein

5. All vitamins and minerals:
   a. provide energy to the body.
   b. serve as parts of body structures.
   c. serve as regulators.
   d. are organic.

6. Characteristics of essential nutrients include:
   a. the body can make them for itself.
   b. they are found in all six classes of nutrients.
   c. they must be received from foods.
   d. a and b
   e. b and c

7. Which of the following is characteristic of elemental diets?
   a. They are sufficient to enable people to thrive.
   b. They support optimal growth and health.
   c. They provide everything that foods provide.
   d. They are life-saving for people in the hospital who cannot eat ordinary food.

8. Within the range set by your genetic inheritance, the likelihood that you will develop chronic diseases is strongly influenced by your food choices.
   a. true
   b. false

9. Which of the following is the most nutrition-responsive disease?

   a. osteoporosis
   b. hypertension
   c. low birthweight
   d. iron-deficiency anemia

10. Which of the following research designs is the most powerful in pinpointing the mechanisms by which nutrition acts because it shows the effects of treatments?

    a. case studies
    b. laboratory studies
    c. epidemiological studies
    d. correlation studies

11. Which of the following is **not** a government agency active in nutrition policy, research and monitoring?

    a. U.S. Department of Health and Human Services
    b. U.S. Department of Agriculture
    c. Food and Drug Administration
    d. Centers for Disease Control and Prevention

12. The number of foods available today may make it more difficult, rather than easier, to plan nutritious diets.

    a. true
    b. false

13. A key to wise diet planning is to make sure your _____ foods are especially nutritious.

    a. processed
    b. fast
    c. enriched
    d. staple

14. Foods composed of parts of whole foods, such as butter, sugar, or corn oil, are called _____ foods.

    a. enriched
    b. partitioned
    c. processed
    d. staple

15. The term that describes a diet that provides no unwanted constituents in excess is:

    a. variety.
    b. calorie control.
    c. moderation.
    d. balance.

16. The key to evaluating an individual food is to:

    a. judge it as either good or bad.
    b. identify how it can reduce your chance of developing an illness.
    c. judge its appropriate role in the context of all other food choices.
    d. listen to what other people say about it.

17. Foods that are rich in nutrients relative to their energy contents are called foods with high _____.

    a. nutrient density
    b. calorie control
    c. nutrient balance
    d. dietary moderation

18. The foods that present the most nutrients per calorie are the:

    a. meats.
    b. vegetables.
    c. fruits.
    d. grains.

19. The most energy-rich of the nutrients is:

    a. carbohydrate.   c. fat.
    b. protein.   d. alcohol.

20. Of the leading causes of death, _____ are related to nutrition, and 1 to alcohol consumption.

    a. 2   c. 4
    b. 3   d. 5

21. Healthy people who eat a healthful diet do not need liquid formula diets and most need no dietary supplements.

    a. true   b. false

22. When a person is fed through a vein:

    a. the digestive organs grow larger.
    b. the body's defenses against infections weaken.
    c. the digestive organs weaken.
    d. a and b
    e. b and c

23. Which of the following is an ongoing national research project that gathers information from about 50,000 people using diet histories, physical examinations and measurements, and laboratory tests?

    a. *Dietary Guidelines for Americans*
    b. National Health and Nutrition Examination Survey
    c. *Healthy People 2010*
    d. Continuing Survey of Food Intakes by Individuals

24. Successes related to *Healthy People 2010* include all of the following except:

    a. reductions in the incidence of foodborne infections.
    b. reductions in death from heart disease.
    c. reductions in cancers.
    d. reductions in the number of overweight people.

25. Nonnutrients affect what characteristic of food?

    a. color.   d. a and b
    b. calories.   e. a and c
    c. taste.

Application Level Items:

26. One of your best friends makes the statement that eggs are a bad food. Which of the following would be the best response to this statement?

    a. Yes they are because they have a lot of cholesterol.
    b. No they are not because they provide high-quality protein.
    c. Yes they are because they cause heart disease.
    d. They are neither good nor bad; their nutritional value depends on how they are used in the total diet.

27. How many calories are contained in a food that has 20 grams of carbohydrate, 15 grams of protein, and 8 grams of fat?

   a. 145
   b. 195
   c. 212
   d. 245

28. A food contains 8 grams of fat and a total of 212 calories. What percentage of the calories in the food come from fat?

   a. 25%
   b. 34%
   c. 46%
   d. 52%

29. An individual consumes a total of 3000 calories and 100 grams of fat in a day. Does this person's diet meet the goal of <35% of calories from fat?

   a. yes
   b. no

30. Susie consumes a diet which allows her to maintain a healthy body weight and to get enough of all the essential nutrients and fiber. She also watches the amounts of sugar, fat and sodium in her diet and eats the same nutritious foods for breakfast each morning. Susie's diet has what characteristic(s) of a nutritious diet?

   a. adequacy
   b. calorie control
   c. variety
   d. a and b
   e. a and c

31. Which of the following characteristics of a nutritious diet is illustrated by the statement that a certain amount of fiber in foods contributes to health of the digestive tract, but too much fiber leads to nutrient losses?

   a. moderation
   b. variety
   c. balance
   d. adequacy

32. Henry has set a goal of increasing his consumption of whole grains and has started eating a high-fiber cereal for breakfast each morning. Henry is currently at which stage of behavior change?

   a. contemplation
   b. preparation
   c. action
   d. maintenance

33. Which of the following changes in a person's diet would be consistent with the *Healthy People 2010* nutrition-related objectives?

   a. eating at least 2 servings of whole grains each day
   b. eating at least 2 servings of fruit each day
   c. eating at least 8 servings of grain products each day
   d. eating at least 1 serving of dark green or orange vegetables each week

34. One cup of vanilla milkshake contains 254 calories and 332 mg calcium; one cup of skim milk has 86 calories and 301 mg calcium. Choosing the skim milk would be an example of applying the principle of:

   a. variety.
   b. calorie control.
   c. moderation.
   d. nutrient density.

Answers

Summing Up

1. food
2. energy
3. six
4. organic
5. water
6. fats
7. minerals
8. essential
9. nutrients
10. growth
11. brain
12. nonnutrients
13. experiments
14. duplicated
15. cultural
16. convenience
17. social
18. nutritional
19. staple
20. *Healthy People*
21. physical activity
22. grains
23. calories
24. adequacy
25. moderation
26. behaviors
27. confidence

Chapter Glossar y

Crossword Puzzle:
1. cuisines
2. nutrient density
3. elemental diets
4. vegetarians
5. omnivores
6. moderation
7. malnutrition
8. chronic diseases
9. balance
10. calories
11. adequacy
12. calorie control
13. variety
14. grams

Matching Exercise:
1. c
2. i
3. l
4. n
5. b
6. f
7. j
8. a
9. h
10. m
11. e
12. k
13. g
14. d
15. o
16. s
17. q
18. p
19. r

Exercises

Chapter Study Questions:

1. Protein can yield energy, but it also provides materials that form structures and working parts of body tissues.

2. Essential nutrients are those that the body cannot make for itself from other raw materials; they are the nutrients that must be obtained in food and are needed by the body; without them you will develop deficiencies.

3. Foods are thought to convey emotional satisfaction and hormonal stimuli that contribute to health; also they provide nonnutrients and other compounds that give them their tastes, aromas, colors, and other characteristics believed to affect health by reducing disease risks.

4. One reason is that some less well-known nutrients and some nonnutrients could be important to health and some foods may be better sources of these than others. Another reason is that a monotonous diet may deliver large amounts of toxins or contaminants. Variety also adds interest and can be a source of pleasure. Finally, a varied diet is likely to be adequate in nutrients.

5. The food industry offers thousands of foods, many of which are processed mixtures of the basic ones, and some of which are constructed mostly from artificial ingredients.

Short Answer Questions:

1. (a) carbohydrate; (b) fat; (c) protein; (d) vitamins

2. (a) water; (b) carbohydrate; (c) fat; (d) protein; (e) vitamins; (f) minerals

3. (a) carbohydrates; (b) fats; (c) protein

4. (a) personal preference; (b) habit; (c) ethnic heritage; (d) social pressure; (e) availability; (f) convenience; (g) economy; (h) positive or negative associations; (i) emotional comfort; (j) values or beliefs; (k) nutritional value; (l) region of the country; (m) weight; (n) advertising

5. (a) adequacy; (b) balance; (c) calorie control; (d) moderation; (e) variety

6. (a) precontemplation; (b) contemplation; (c) preparation; (d) action; (e) maintenance; (f) adoption/moving on

Problem-Solving:

1. b 2. e 3. c 4. f 5. a 6. g 7. d

Controversy Questions:

1. The work must survive a screening review by peers of scientists.

2. Membership in the American Society for Clinical Nutrition.

3. A fraudulent business that awards meaningless degrees or certificates of competency to those who pay the fees, without requiring its students to meet educational standards.

4. Completion of a two-year academic degree from an accredited college or university and an approved dietetic technician program.

5. Administrative, clinical and nutrition support.

6. The National Library of Medicine's PubMed website; it provides free access to over 10 million abstracts of research papers published in scientific journals.

Sample Test Items

1. c (p. 13)
2. a (p. 6)
3. d (p. 6)
4. b (pp. 6-7)
5. c (p. 7)
6. e (p. 7)
7. d (pp. 7-8)
8. a (p. 3)
9. d (p. 3)
10. b (p. 14)
11. c (p. 16)
12. a (p. 8)
13. d (p. 9)
14. b (p. 9)
15. c (pp. 10-11)
16. c (p. 9)
17. a (p. 20)
18. b (p. 20)
19. c (p. 7)
20. c (p. 3)
21. a (p. 7)
22. e (p. 8)
23. b (p. 16-17)
24. d (p. 6)
25. e (p. 8)
26. d (p. 9)
27. c (p. 7)
28. b (p. 7)
29. a (p. 7)
30. d (pp. 9-11)
31. a (p. 11)
32. b (p. 19)
33. b (p. 5)
34. d (p. 20)

Chapter 2 - Nutrition Tools
Standards and Guidelines

Chapter Objectives

After completing this chapter, you should be able to:

1. Describe the goals of the Dietary Reference Intakes (DRI) Committee and the purposes of the nutrient intake standards—RDA, AI, EAR, UL and AMDR.

2. Explain how Dietary Reference Intake values are established for nutrients.

3. Recognize the purpose of Daily Values and describe their strengths and limitations.

4. List the recommendations of the *Dietary Guidelines for Americans*.

5. Describe the components of the USDA Food Guide and utilize it to plan a nutritious diet.

6. Explain how exchange systems are useful to diet planners.

7. Discuss how consumers can use the ingredients list, Nutrition Facts panel, and health messages on food labels to make healthy food choices.

8. Describe possible actions of phytochemicals and functional foods and the evidence supporting their role in fighting certain diseases (Controversy 2).

Key Concepts

✓ The Dietary Reference Intakes are nutrient intake standards set for people living in the United States and Canada. The Daily Values are U.S. standards used on food labels.

✓ The DRI provide nutrient intake goals for individuals, provide a set of standards for researchers and makers of public policy, establish tolerable upper limits for nutrients that can be toxic in excess, and take into account evidence from research on disease prevention. The DRI are composed of the RDA, AI, UL and EAR lists of values along with the AMDR ranges for energy-yielding nutrients.

✓ The DRI represent up-to-date, optimal, and safe nutrient intakes for healthy people in the United States and Canada.

✓ The DRI are based on scientific data and are designed to cover the needs of virtually all healthy people in the United States and Canada.

✓ Estimated Energy Requirements are energy intake recommendations predicted to maintain body weight and to discourage unhealthy weight gain.

✓ The Daily Values are standards used only on food labels to enable consumers to compare the nutrient values among foods.

✓ The *Dietary Guidelines for Americans, Nutrition Recommendations for Canadians*, and other recommendations address the problems of overnutrition and undernutrition. To implement them requires exercising regularly, following the USDA Food Guide, seeking out whole grains, fruits, and vegetables, limiting intakes of saturated and *trans* fats, sugar, and salt, and moderating alcohol intake.

✓ The USDA Food Guide specifies the amounts of foods from each group people need to consume to meet their nutrient requirements without exceeding their calorie allowances.

✓ The concepts of the USDA Food Guide are conveyed to consumers through the MyPyramid educational tool.

✓ The USDA Food Guide can be used with flexibility by people with different eating styles.

✓ People wishing to avoid overconsuming calories must pay attention to the sizes of their food servings.

✓ Exchange lists facilitate calorie control by providing an understanding of how much carbohydrate, fat, and protein are in each food group.

Summing Up

A committee of qualified nutrition experts from the United States and (1)_____ develops and publishes nutrient intake recommendations known as the (2)_____. Most people need to focus on only two kinds of DRI values: those that set nutrient intake goals for individuals, the Recommended Dietary Allowances (RDA) and (3)_____; and those that define an upper limit of safety for nutrient intakes (UL). These recommendations are intended to be used to plan diets for (4)_____. Whenever there is insufficient evidence to generate an RDA, the DRI Committee establishes an (5)_____ value instead. Other values established by the DRI Committee include (6)_____, which are used by researchers and policy makers, and Tolerable Upper Intake Levels. The value for energy intake is set at a level predicted to maintain (7)_____. Healthy ranges of intake for each energy nutrient, known as the (8)_____, have also been established. The DRI are designed for health maintenance and (9)_____ in healthy people.

The purpose of the *Dietary Guidelines for Americans* is to promote health and reduce chronic diseases through diet and (10)_____. The USDA Food Guide is a food group plan to help people achieve the goals of the (11)_____. The USDA Food Guide sorts foods in each group by (12)_____ and includes (13)_____ food groups. The concept of the (14)_____—the difference between the calories needed to maintain weight and those needed to supply nutrients—gives people the option to consume some less nutrient-dense foods while controlling (15)_____.

Exchange systems are useful for those wishing to control (16)_____, for those who must control carbohydrate intakes, and those who should control their intake of (17)_____ and saturated fat. The lists organize foods according to their calorie, carbohydrate, fat, saturated fat and

(18)_____ contents. The exchange system highlights a fact that the USDA Food Guide also points out, that most foods provide more than just one (19)_____ nutrient.

The Nutrition Education and Labeling Act of 1990 set the requirements for nutrition labeling information. According to law, every packaged food must state the (20)_____ name of the product, the name and address of the manufacturer, the (21)_____ contents in terms of weight and the ingredients, in (22)_____ order of predominance by weight. The (23)_____ informs consumers of the nutrient contents of the food. In addition to food energy, the label must include total fat (with breakdown of saturated fat and *trans* fat), cholesterol, sodium, total carbohydrate (including starch, fiber and sugars), (24)_____, vitamins A and C, and the minerals (25)_____ and iron. Some claims about health are allowed on food labels and these are regulated by the (26)_____.

Chapter Glossary

Matching Exercise:

_____ 1. requirement

_____ 2. discretionary calorie allowance

_____ 3. food group plans

_____ 4. Daily Values

_____ 5. MyPyramid

_____ 6. balance study

_____ 7. exchange system

_____ 8. Estimated Energy Requirement

_____ 9. Dietary Reference Intakes

a. nutrient standards that are printed on food labels

b. the amount of a nutrient that will just prevent the development of specific deficiency signs

c. diet-planning tools that sort foods into groups based on nutrient content and then specify that people should eat certain minimum numbers of servings of foods from each group

d. a laboratory study in which a person is fed a controlled diet and the intake and excretion of a nutrient are measured

e. a diet planning tool that organizes foods with respect to their nutrient contents and calorie amounts

f. a set of four lists of values for measuring the nutrient intakes of healthy people in the United States and Canada

g. the average dietary energy intake predicted to maintain energy balance in a healthy adult of a certain age, gender, weight, height, and level of physical activity consistent with good health

h. the balance of calories remaining in a person's energy allowance after accounting for the number of calories needed to meet recommended nutrient intakes through consumption of nutrient-dense foods

i. an educational tool that promotes taking small steps each day towards healthy changes in diet and lifestyle

Exercises

Answer these chapter study questions:

1. Why is the value set for energy intake not generous?

2. Describe the goals the DRI committee had in mind when setting the DRI values.

3. List the foods most Americans need to choose more often and less often in order to meet the ideals of the 2005 *Dietar yGuidelines* .

4. Explain how the concept of the discretionary calorie allowance may be used to control intake of calories while meeting nutrient needs.

5. Explain why the DRI values are not used on food labels.

6. How do USDA Food Guide serving equivalents compare to the portion sizes served in restaurants?

7. Describe the current system the FDA uses for health claims on food labels.

Complete these short answer questions:

1. The exchange system lists estimates of these nutrients in foods:

    a.   c.

    b.   d.

2. The UL are indispensable to consumers who:

    a.

    b.

3. Dietary Reference Intakes include these five lists of values:

    a.

    b.

    c.

    d.

    e.

4. The five groups in the USDA Food Guide include:

    a.

    b.

    c.

    d.

    e.

5. The subgroups within the vegetable group in the USDA Food Guide include:

   a.

   b.

   c.

   d.

   e.

6. Food labels must provide the following information on the panel called "Nutrition Facts":

   a.

   b.

   c.

   d.

   e.

   f.

   g.

   h.

   i.

   j.

   k.

   l.

7. List 5 ways someone may choose to "spend" their discretionary calorie allowance:

   a.

   b.

   c.

   d.

   e.

Solve these problems:

1. Use the menu below to complete sections a-d which follow:

Menu		
Breakfast	**Lunch**	**Dinner**
1 hard-boiled egg	2 tbs. peanut butter	3 oz. hamburger
1 cup cornflakes	1 tsp. jelly	hamburger bun
1 cup skim milk	2 slices bread	1 tbs. low-fat mayonnaise
1 medium grapefruit	1 medium apple	½ cup green beans
	1 diet cola	1 cup plain, nonfat yogurt

a. Categorize the foods in the menu according to the USDA Food Guide:

Fruits	Vegetables	Grains	Meat, Poultry, Fish, Dried Beans, Eggs, and Nuts	Milk, Yogurt, and Cheese	Oils

b. Which foods were you unable to classify?

c. Evaluate the adequacy of the menu according to the USDA Food Guide by comparing the actual amounts with the recommended amounts of the following for a sedentary woman 32 years old with a daily allowance of 1,800 calories (see Table 2-3 on page 42).

Food Group	Recommended Amount	Actual Amount
Fruits	1 ½ cups	_____
Vegetables	2 ½ cups	_____
Grains	6 oz.	_____
Meat, poultry, fish, dried beans, eggs, and nuts	5 oz.	_____
Milk, yogurt, and cheese	3 cups	_____
Oils	5 tsp.	_____

d. Categorize the foods in the menu into exchanges by placing an X in the appropriate blank. (See p. 48 for an explanation of how foods are classified into exchange lists.)

	Starch	Meats	Vegetables	Fruits	Milks	Fats	Other
egg	___	___	___	___	___	___	___
cornflakes	___	___	___	___	___	___	___
skim milk	___	___	___	___	___	___	___
jelly	___	___	___	___	___	___	___
grapefruit	___	___	___	___	___	___	___
peanut butter	___	___	___	___	___	___	___
bread	___	___	___	___	___	___	___
apple	___	___	___	___	___	___	___
diet cola	___	___	___	___	___	___	___
hamburger	___	___	___	___	___	___	___
bun	___	___	___	___	___	___	___
mayonnaise	___	___	___	___	___	___	___
green beans	___	___	___	___	___	___	___
plain yogurt	___	___	___	___	___	___	___

2. a. How many calories are in a day's meal which provides 260 grams carbohydrate, 30 grams protein, and 60 grams of fat? _____ calories

   b. What percentage of the calories is from carbohydrate? _____ %

   c. What percentage of the calories is from fat? _____ %

   d. How does the percentage of calories from fat compare to the amount in the Acceptable Macronutrient Distribution Ranges specified by the DRI Committee?

Answer these controversy questions:

1. What advice would you give someone who looks to consume chocolate to reduce the risk of heart attack and stroke?

2. Why do food manufacturers often refine away the natural flavonoids in foods?

3. What are some foods common to many Asian diets that contain phytoestrogens?

4. Which types of cancer occur less frequently in people who consume five tomato-containing meals per week, and why?

5. Describe arguments against the use of phytochemical supplements.

Study Aids

Use Tables 2-3 and 2-4 on page 42 and Figure 2-5 on pages 38-39 to study the USDA Food Guide, including the food groups, examples of foods in each group, nutrient density of foods in each group, and amounts of food recommended for specific target groups.

Sample Test Items

Comprehension Level Items:

1. Recommendations for dietary nutrient intakes of healthy people in the United States and Canada are called:

   a. Recommended Nutrient Intakes.
   b. Daily Values.
   c. Adequate Intakes.
   d. Dietary Reference Intakes.
   e. Tolerable Upper Intake Levels.

2. Which of the following establishes nutrient requirements for given life stages and gender groups that researchers and policy makers use in their work?

   a. Adequate Intakes
   b. Tolerable Upper Intake Limits
   c. Estimated Average Requirements
   d. Recommended Dietary Allowances

3. All of the following adjectives are descriptive of the Dietary Reference Intakes except:

   a. minimal.
   b. optimal.
   c. safe.
   d. up-to-date.

4. The value set for energy, the Estimated Energy Requirements, is not generous in order to:

   a. compensate for lack of physical activity.
   b. discourage overconsumption of energy which would lead to obesity.
   c. encourage a decrease in fat consumption.
   d. ensure generous allowances of fat and carbohydrate.

5. The term *Daily Values* appears on:

   a. fresh meats.
   b. refined foods.
   c. food labels.
   d. fresh vegetables.

6. Which of the following could not be part of a person's discretionary calorie allowance?

   a. calories from added sugars, such as jams, colas and honey
   b. calories from nutrient-dense foods needed to meet recommended nutrient intakes
   c. calories from added fat absorbed by the batter on a fried chicken leg
   d. calories from the naturally occurring fat in whole milk, which is consumed instead of nonfat milk

7. The USDA Food Guide recommends intakes of certain amounts from each food group based on age alone.

   a. true
   b. false

8. Exchange systems are particularly helpful for:

   a. weight watchers.
   b. people with diabetes.
   c. people with hypertension.
   d. a and b
   e. b and c

9. Which values do public health officials use to set safe upper limits for nutrients added to the food and water supply?

   a. Adequate Intakes
   b. Tolerable Upper Intake Levels
   c. Recommended Dietary Allowances
   d. Estimated Average Requirements

10. The *Dietary Guidelines for Americans* apply to most people aged two or older.

    a. true
    b. false

11. Corn and potatoes are listed with _____ in the exchange system.

    a. meats
    b. vegetables
    c. breads
    d. fats

12. The exchange system highlights the fact that most foods provide more than one energy nutrient.

    a. true
    b. false

13. Based on the USDA Food Guide, which of the following would be recommended as primary sources of fats in the diet?

    a. fat that occurs naturally in milk, butter, sour cream
    b. fat that occurs naturally in the skin of poultry, fat streaks in beef, cream cheese
    c. lard, margarine, pudding made from whole milk
    d. olive oil, peanut oil, oils that occur naturally in fatty fish

14. The 2005 *Dietary Guidelines* advise Americans to increase consumption of all of the following **except**:

    a. refined grains.
    b. vegetables.
    c. fruits.
    d. fat-free or low-fat milk or milk products.

15. The American College of Sports Medicine recommends exercise for a duration of at least 30 minutes total daily.

    a. true
    b. false

16. Which of the following is the major purpose of the Daily Values listed on food labels?

    a. prevent chronic diseases
    b. allow comparisons among foods
    c. discourage overconsumption of calories
    d. prevent nutrient toxicity

17. The calculations used to determine the "% Daily Value" figures for nutrient contributions from a serving of food are based on a 2,000- to 2,500-calorie diet.

    a. true
    b. false

18. Which of the following nutrient information is **not** required on food labels?

    a. fat in grams per serving
    b. polyunsaturated fat in grams per serving
    c. saturated fat in grams per serving
    d. cholesterol in milligrams per serving

19. Which of the following FDA grades for health claims, relating the roles of nutrients and food constituents to disease states, represents significant scientific agreement?

   a. A
   b. B
   c. C
   d. D

20. Until recently, FDA held manufacturers to the highest standards of scientific evidence before allowing them to place health claims on food labels.

   a. true
   b. false

21. The absence of a Tolerable Upper Intake Level for a nutrient does not imply that it is safe to consume it in any amount.

   a. true
   b. false

22. The DRI Committee recommends a diet that contains _____ % of its calories from fat.

   a. 10-35
   b. 20-35
   c. 45-65
   d. 50-70

23. The DRI are **not** designed for:

   a. disease prevention.
   b. health restoration.
   c. repletion of nutrients.
   d. a and b
   e. b and c

24. Which of the following meals represents the **least** nutrient-dense Mexican food choices?

   a. corn tortilla, pinto beans, tomatoes, guava
   b. flour tortilla, black beans, iceberg lettuce, pineapple
   c. fried tortilla shell, refried beans, cheddar cheese, avocado
   d. lean chicken breast, squash, tomatoes, corn, mango

25. The USDA Food Guide can assist vegetarians in their food choices.

   a. true
   b. false

Application Level Items:

26. Someone consumes a 2,000-calorie diet composed of 150 grams carbohydrate, 60 grams protein, and 128 grams fat. Does this person meet the Acceptable Macronutrient Distribution Range for protein established by the DRI Committee?

   a. yes
   b. no

27. Does the diet described in question #26 meet the Acceptable Macronutrient Distribution Range for fat?

   a. yes
   b. no

28. An ingredient list states that a food product contains peas, water, carrots, sugar, and salt. Which of these ingredients is present in the smallest quantity?

a. peas
b. carrots
c. sugar
d. salt
e. water

29. You see the following ingredient list on a cereal label: wheat, sugar, raisins, brown sugar, syrup, and salt. You therefore conclude that this product:

a. is a good source of fiber.
b. contains 100% whole wheat.
c. is a nutritious food choice.
d. contains close to half its weight in sugar.

30. The Nutrition Facts panel on a box of cereal lists the following information for amounts per serving: 167 calories; 27 calories from fat; 3 g total fat. What percentage of the calories are provided by fat?

a. 3%
b. 11%
c. 16%
d. 27%

31. Stan wants to apply the concept of the discretionary calorie allowance to his diet. If he wants to attend his niece's birthday party and eat cake and ice cream tomorrow evening, what should he do?

a. Stan should eat low-calorie, nutrient-dense foods from each food group during the day tomorrow, so that he will have enough discretionary calories left to spend on cake and ice cream after his nutrient needs are met.
b. Stan should fast during the day so that he will have plenty of discretionary calories to spend on cake and ice cream at the party.
c. Stan should not eat cake and ice cream at the party; only nutrient-dense foods may be eaten, and cake and ice cream are energy-dense but not nutrient-dense.
d. Stan should eat whatever he wants tomorrow but keep track of the calories; it doesn't matter where his calories come from, as long as he doesn't consume more calories than his body uses.

32. Cereal A contains 2 grams of fat and 120 calories per serving. Cereal B contains 3 grams of fat and 230 calories per serving. Which cereal could be labeled as low fat?

a. cereal A
b. cereal B
c. both
d. neither

Answers

Summing Up

1. Canada
2. Dietary Reference Intakes
3. Adequate Intakes
4. individuals
5. Adequate Intake
6 Estimated Average Requirements
7. body weight
8. Acceptable Macronutrient Distribution Ranges
9. disease prevention
10. physical activity
11. *Dietary Guidelines*
12. nutrient density
13. five
14. discretionary calorie allowance
15. calories
16. calories
17. fat
18. protein
19. energy
20. common
21. net
22. descending
23. Nutrition Facts panel
24. protein
25. calcium
26. Food and Drug Administration

Chapter Glossary

Matching Exercise:

1. b	3. c	5. i	7. e	9. f
2. h	4. a	6. d	8. g	

Exercises

Chapter Study Questions:

1. Because too much energy causes unhealthy weight gain and associated diseases.

2. The four goals of the committee include setting intake recommendations for individuals, facilitating nutrition research and policy, establishing safety guidelines, and preventing chronic diseases.

3. For most people, meeting the diet ideals of the *Dietary Guidelines* requires choosing more vegetables (especially dark green vegetables, orange vegetables, and legumes), fruits, whole grains and fat-free or low-fat milk and milk products. It also requires choosing less refined grains, total fats (especially saturated fat, *trans* fat, and cholesterol), added sugars, and total calories.

4. The discretionary calorie allowance is the balance of calories remaining in a person's energy allowance after accounting for the number of calories needed to meet recommended nutrient intakes through consumption of nutrient-dense foods. Any calories in a food which are not necessary in order to obtain its nutrients, including added fats, sugars, and naturally occurring fats in higher-fat versions of foods (such as whole milk, which could be replaced by nonfat milk), are considered discretionary. To apply this concept, a person must plan their diet by choosing nutrient-dense foods first, ensuring adequacy; once nutrient needs are met, any additional calories left in their allowance may be consumed as the person wishes, or omitted (to lose weight).

5. The DRI values vary from group to group, whereas on a food label, one set of values must apply to everyone. The Daily Values reflect the needs of an "average person."

6. USDA serving equivalents are specific, precise, and reliable for delivering certain amounts of key nutrients in foods. In contrast, portion sizes served in restaurants tend to be much larger than USDA serving equivalents in order to ensure repeat business.

7. In the past, FDA held manufacturers to the highest standards of scientific evidence before allowing them to place health claims on food labels. Now the FDA allows other claims supported by weaker evidence and assigns each claim a letter grade reflecting the degree to which the claim is backed by science.

Short Answer Questions:

1. (a) carbohydrate; (b) fat; (c) saturated fat; (d) protein

2. (a) take supplements; (b) consume foods and beverages to which vitamins or minerals are added

3. (a) Recommended Dietary Allowances; (b) Adequate Intakes; (c) Tolerable Upper Intake Levels; (d) Estimated Average Requirements; (e) Acceptable Macronutrient Distribution Ranges

4. (a) fruits; (b) vegetables; (c) grains; (d) meat, poultry, fish, dried peas and beans, eggs and nuts; (e) milk, yogurt and cheese

5. (a) dark green vegetables; (b) orange and deep yellow vegetables; (c) legumes; (d) starchy vegetables; (e) other vegetables

6. (a) serving size; (b) servings per container; (c) total calories and calories from fat; (d) fat grams per serving including saturated fat and *trans* fat; (e) cholesterol; (f) sodium; (g) total carbohydrate including fiber, starch and sugars; (h) protein; (i) vitamin A; (j) vitamin C; (k) calcium; (l) iron

7. (a) extra portions of nutrient-dense foods; (b) naturally-occurring or added fats; (c) added sugars; (d) alcohol; (e) omitting them from the diet

Problem-Solving:

1. a.

Fruits	Vegetables	Grains	Meat, Poultry, Fish, Dried Beans, Eggs, and Nuts	Milk, Yogurt, and Cheese	Oils
grapefruit apple	green beans	cornflakes bread hamburger bun	egg peanut butter hamburger	skim milk yogurt	mayonnaise

b. jelly; diet cola

c. fruits – 1 cup; vegetables – ½ cup; grains – 5 oz.; meat, poultry, fish, dried beans, eggs, and nuts – 6 oz.; milk, yogurt, and cheese – 2 cups; oils – 1 tsp.

This menu does not provide enough fruits, vegetables, whole grains, non-fat dairy products, or healthful oils for the woman described.

d.

	Starch	Meats	Vegetables	Fruits	Milks	Fats	Other
egg		X					
cornflakes	X						
skim milk					X		
jelly							X
grapefruit				X			
peanut butter		X					
bread	X						
apple				X			
diet cola							X
hamburger		X					
bun	X						
mayonnaise						X	
green beans			X				
plain yogurt					X		

2. (a) 1700 calories; (b) 61% calories from carbohydrate; (c) 32% calories from fat; (d) the percent calories from fat is within the recommended level of 20-35%

Controversy Questions:

1. It would be better to obtain flavonoids from nutrient-dense food sources such as fruits, vegetables or green or black tea because chocolate promotes weight gain and contributes fat and sugar to the diet. However, chocolate can be used as an occasional treat within the context of the dietary principles of moderation, variety, and balance.

2. Because they impart a bitter taste to foods and most consumers desire milder flavors.

3. Soybeans and their products including tofu, miso and soy drink.

4. Cancers of the esophagus, prostate and stomach; because tomatoes contain lycopene, which has antioxidant activity and may inhibit reproduction of cancer cells.

5. The body is able to handle phytochemicals in natural foods because they are diluted among all of the other constituents. The body is not adapted to the concentrated doses of phytochemicals in supplements. There is also a lack of evidence for the safety of isolated phytochemicals in humans and the safety and effectiveness of these supplements are not adequately regulated.

Sample Test Items

1. d (p. 30)
2. c (p. 31)
3. a (pp. 30-35)
4. b (pp. 34-35)
5. c (p. 35)
6. b (p. 41)
7. b (p. 42)
8. d (p. 48)
9. b (p. 31)
10. a (p. 35)
11. c (p. 48)
12. a (p. 48)
13. d (p. 39)
14. a (p. 37)
15. a (p. 37)
16. b (p. 35)
17. a (p. 35)
18. b (p. 50)
19. a (p. 53)
20. a (p. 52)
21. a (p. 31)
22. b (p. 31)
23. e (p. 33)
24. c (p. 47)
25. a (p. 44)
26. a (p. 31)
27. b (p. 31)
28. d (pp. 50-51)
29. d (pp. 50-51)
30. c (pp. 49-52)
31. a (p. 41)
32. c (p. 51)

Chapter 3 - The Remarkable Body

Chapter Objectives

After completing this chapter, you should be able to:

1. Identify the basic needs of cells and describe how cells are organized into tissues, organs, and systems.

2. Describe the major function of the cardiovascular system and how it ensures that the body's fluids circulate properly among all organs.

3. Explain how nutrition affects the hormonal system and how hormones affect nutrition.

4. Describe the nervous system's role in hunger regulation.

5. Discuss how the immune system functions to enable the body to resist diseases.

6. Explain the mechanical and chemical digestive processes in order of their occurrence in the body.

7. Describe how the body absorbs and transports nutrients.

8. Identify the body's organs which play roles in removing wastes from the body.

9. Describe the body's major storage systems for nutrients.

10. Describe alcohol's effects on the body and evaluate whether the benefits of alcohol outweigh the risks (Controversy 3).

Key Concepts

✓ The body's cells need energy, oxygen, and nutrients, including water, to remain healthy and do their work. Genes direct the making of each cell's machinery, including enzymes. Genes and nutrients interact in ways that affect health. Specialized cells are grouped together to form tissues and organs; organs work together in body systems.

✓ Blood and lymph deliver nutrients to all the body's cells and carry waste materials away from them. Blood also delivers oxygen to cells. The cardiovascular system ensures that these fluids circulate properly among all organs.

✓ Glands secrete hormones that act as messengers to help regulate body processes.

✓ The nervous system joins the hormonal system to regulate body processes through communication among all the organs. Together, the hormonal and nervous systems respond to the need for food, govern the act of eating, regulate digestion, and call for the stress response.

✓ A properly functioning immune system enables the body to resist diseases.

✓ The preference for sweet, salty, and fatty tastes seems to be inborn and can lead to overconsumption of foods that offer them.

✓ The digestive tract is a flexible, muscular tube that digests food and absorbs its nutrients and some nonnutrients. Ancillary digestive organs aid digestion.

✓ The digestive tract moves food through its various processing chambers by mechanical means. The mechanical actions include chewing, mixing by the stomach, adding fluid, and moving the tract's contents by peristalsis. After digestion and absorption, wastes are excreted.

✓ Chemical digestion begins in the mouth, where food is mixed with an enzyme in saliva that acts on carbohydrates. Digestion continues in the stomach, where stomach enzymes and acid break down protein. Digestion then continues in the small intestine; there the liver and gallbladder contribute bile that emulsifies fat, and the pancreas and small intestine donate enzymes that continue digestion so that absorption can occur. Bacteria in the colon break down certain fibers.

✓ The healthy digestive system is capable of adjusting to almost any diet and can handle any combination of foods with ease.

✓ The mechanical and chemical actions of the digestive tract break foods down to nutrients, and large nutrients to their smaller building blocks, with remarkable efficiency.

✓ The digestive system feeds the rest of the body and is itself sensitive to malnutrition. The folds and villi of the small intestine enlarge its surface area to facilitate nutrient absorption through countless cells to the blood and lymph. These transport systems then deliver the nutrients to all the body's cells.

✓ The digestive tract has many ways to communicate its needs. By taking the time to listen, you will obtain a complete understanding of the mechanics of the digestive tract and its signals.

✓ The kidneys adjust the blood's composition in response to the body's needs, disposing of everyday wastes and helping remove toxins. Nutrients, including water, and exercise help keep the kidneys healthy.

✓ The body's energy stores are of two principal kinds: glycogen in muscle and liver cells (in limited quantities) and fat in fat cells (in potentially large quantities). Other tissues store other nutrients.

✓ To achieve optimal function, the body's systems require nutrients from outside. These have to be supplied through a human being's conscious food choices.

Summing Up

The body is composed of trillions of (1)_____ and the nature of their work is determined by genes. Among the cells' most basic needs are (2)_____ and the oxygen with which to burn it, and nutrients to remain healthy and do their work. Specialized cells are grouped together to form tissues and (3)_____. Body (4)_____ supply the tissues continuously with energy, oxygen and nutrients. The body's main fluids are the blood and (5)_____. Blood travels within the arteries, (6)_____ and capillaries, as well as within the heart's chambers.

As the blood travels through the cardiovascular system, it picks up oxygen in the (7)_____ and also releases carbon dioxide there. As it passes through the digestive system, the blood picks up most (8)_____, other than fats, for distribution elsewhere. All blood leaving the digestive system is routed directly to the (9)_____. The blood is cleansed of its wastes as it passes through the (10)_____.

The hormonal and (11)_____ systems regulate body processes through communication among all the organs. They respond to the need for food, govern the act of eating, regulate (12)_____ and call for the stress response. The (13)_____ system enables the body to resist diseases. The first type of white blood cells to defend the body against invaders are called (14)_____. the (15)_____ known as T-cells and B-cells, are other major players in the body's immune system. The (16)_____ system's job is to digest food to its component nutrients and then to (17)_____ those nutrients, leaving behind the substances that are appropriate to excrete. To do this, the system works at two levels, one mechanical and the other (18)_____.

The body's cells need nutrients around the (19)_____ which requires systems of (20)_____ and release to meet the cells' needs between meals. However, some (21)_____ are stored in the body in much larger quantities than others. Body tissues store excess energy-containing nutrients in two forms, glycogen and (22)_____. Body stores for other nutrients include the liver and fat cells which store many vitamins and the (23)_____ which provide reserves of calcium and other minerals.

Chapter Glossary

Matching Exercise 1:

_____ 1. antibodies

_____ 2. metabolism

_____ 3. immune system

_____ 4. microbes

_____ 5. antigen

_____ 6. lymphocytes

_____ 7. pancreas

_____ 8. neurotransmitters

_____ 9. microvilli

_____ 10. phagocytes

_____ 11. hiccups

_____ 12. T-cells

_____ 13. irritable bowel syndrome

_____ 14. ulcer

_____ 15. B-cells

_____ 16. gastroesophageal reflux disease

_____ 17. antacids

_____ 18. glycogen

a. lymphocytes that attack antigens

b. a system of tissues and organs that defend the body against antigens, foreign materials that have penetrated the skin or body linings

c. the sum of all physical and chemical changes taking place in living cells

d. intermittent disturbance of bowel function, especially diarrhea or alternating diarrhea and constipation

e. chemicals that are released at the end of a nerve cell when a nerve impulse arrives

f. proteins, made by cells of the immune system, that are expressly designed to combine with and to inactivate specific antigens

g. bacteria, viruses, or other organisms invisible to the naked eye, some of which cause diseases

h. medications that react directly and immediately with the acid of the stomach, neutralizing it

i. spasms of both the vocal cords and the diaphragm, causing periodic, audible, short, inhaled coughs

j. a microbe or substance that is foreign to the body

k. white blood cells that can ingest and destroy antigens

l. white blood cells that participate in the immune response; B-cells and T-cells

m. lymphocytes that produce antibodies

n. a severe and chronic splashing of stomach acid and enzymes into the esophagus, throat, mouth, or airway that causes inflammation and injury to those organs

o. a storage form of carbohydrate energy

p. tiny, hairlike projections on each cell of every villus that can trap nutrient particles and absorb them into the cells

q. an organ with two main functions; one is an endocrine function and the other is an exocrine function

r. an erosion in the topmost, and sometimes underlying, layers of cells that form a lining

Matching Exercise 2:

_____ 1. kidneys

_____ 2. liver

_____ 3. digest

_____ 4. cortex

_____ 5. intestine

_____ 6. norepinephrine

_____ 7. veins

_____ 8. digestive system

_____ 9. blood

_____ 10. large intestine

_____ 11. heartburn

_____ 12. pyloric valve

_____ 13. hernia

_____ 14. constipation

_____ 15. small intestine

_____ 16. sphincter

_____ 17. bladder

a. to break molecules into smaller molecules; a main function of the digestive tract with respect to food

b. the fluid of the cardiovascular system composed of water, red and white blood cells, other formed particles, nutrients, oxygen, and other constituents

c. the outermost layer of something

d. blood vessels that carry blood, with the carbon dioxide it has collected, from the tissues back to the heart

e. a compound related to epinephrine that helps to elicit the stress response

f. a pair of organs that filter wastes from the blood, make urine, and release it to the bladder for excretion from the body

g. a large, lobed organ that lies just under the ribs

h. a long, tubular organ of digestion and the site of nutrient absorption

i. the body system composed of organs that break down complex food particles into smaller, absorbable products

j. the portion of the intestine that completes the absorption process

k. the circular muscle of the lower stomach that regulates the flow of partly digested food into the small intestine

l. a 20-foot length of small-diameter intestine that is the major site of digestion of food and absorption of nutrients

m. infrequent, difficult bowel movements often caused by diet, inactivity, dehydration, or medication

n. a protrusion of an organ or part of an organ through the wall of the body chamber that normally contains the organ

o. a burning sensation in the chest (heart) area caused by backflow of stomach acid into the esophagus

p. a circular muscle surrounding, and able to close, a body opening

q. the sac that holds urine until time for elimination

Crossword Puzzle 1:

Word Bank:

absorb	enzyme	insulin
adipose tissue	epinephrine	organs
arteries	fat cells	peristalsis
body system	glucagon	tissues
capillaries	hormones	
cells	hypothalamus	

Crossword Puzzle 1 Across:

4. the body's fat tissue, consisting of masses of fat-storing cells and blood vessels to nourish them
5. the wave-like muscular squeezing of the esophagus, stomach, and small intestine that pushes their contents along
6. discrete structural units made of tissues that perform specific jobs, such as the heart, liver, and brain
7. minute, weblike blood vessels that connect arteries to veins and permit transfer of materials between blood and tissues
10. a hormone from the pancreas that helps glucose enter cells from the blood
12. a group of related organs that work together to perform a function
13. chemicals that are secreted by glands into the blood in response to conditions in the body that require regulation
14. blood vessels that carry blood containing fresh oxygen supplies from the heart to the tissues
15. systems of cells working together to perform specialized tasks
16. cells that specialize in the storage of fat and that form the fat tissue

Crossword Puzzle 1 Down:

1. a part of the brain that senses a variety of conditions in the blood, such as temperature, glucose content, salt content, and others
2. a hormone from the pancreas that stimulates the liver to release glucose into the bloodstream
3. the smallest units in which independent life can exist
8. the major hormone that elicits the stress response
9. a working protein that speeds up a specific chemical reaction
11. to take in, as nutrients are taken into the intestinal cells after digestion; the main function of the digestive tract with respect to nutrients

Crossword Puzzle 2:

Crossword Puzzle 2 Across:

1. muscular, elastic, pouchlike organ of the digestive tract that grinds and churns swallowed food and mixes it with acid and enzymes, forming chyme
6. the fluid resulting from the actions of the stomach upon a meal
7. the fluid that moves from the bloodstream into tissue spaces and then travels in its own vessels, which eventually drain back into the bloodstream
10. the cell-free fluid part of blood and lymph
11. fluid residing outside the cells that transports materials to and from cells
12. the body's organs of gas exchange
13. the large intestine
15. frequent, watery bowel movements usually caused by diet, stress, or irritation of the colon
16. waste material remaining after digestion and absorption are complete

Crossword Puzzle 2 Down:

2. a slippery coating of the digestive tract lining (and other body linings) that protects the cells from exposure to digestive juices (and other destructive agents)
3. fingerlike projections of the sheets of cells that line the intestinal tract
4. a compound with both water-soluble and fat-soluble portions that can attract fats and oils into water, combining them
5. the working units in the kidneys, consisting of intermeshed blood vessels and tubules
8. a cholesterol-containing digestion fluid made by the liver, stored in the gallbladder, and released into the small intestine when needed
9. a common alkaline chemical; a secretion of the pancreas
14. a measure of acidity on a point scale

Word Bank:

bicarbonate	diarrhea	lungs	pH
bile	emulsifier	lymph	plasma
chyme	extracellular fluid	mucus	stomach
colon	feces	nephrons	villi

Exercises

Answer these chapter study questions:

1. Identify and describe three factors necessary to ensure efficient circulation of fluid to all body cells.

2. Differentiate between the functions of glands and hormones.

3. Differentiate between the mechanical aspect of digestion and the chemical aspect of digestion.

4. Describe how the body's digestive system is affected by:

   a. severe undernutrition.

   b. lack of dietary fiber.

5. What is the difference between glycogen and fat?

6. Why do health care professionals inspect the skin and inside of the mouth to detect signs of malnutrition?

Complete these short answer questions:

1. The cells' most basic needs include:

   a.   c.

   b.

2. The body's main fluids are the _____ and _____.

3. The central controllers of the nervous system are:

   a.   b.

4. The body's hormonal balance is altered by:

   a.   c.

   b.

5. Five factors required to support the health of the kidneys include:

   a.

   b.

   c.

   d.

   e.

6. The liver converts excess energy containing nutrients into two forms including _____ and
   _____.

Solve these problems:

For questions 1-8, match the organs of the digestive system, listed on the left, to their primary functions, listed on the right.

_____ 1. stomach	a.	manufactures enzymes to digest all energy-producing nutrients and releases bicarbonate to neutralize stomach acid
_____ 2. liver	b.	stores bile until needed
_____ 3. large intestine	c.	churns, mixes, and grinds food to a liquid mass
_____ 4. mouth	d.	passes food to the stomach
	e.	manufactures bile to help digest fats
_____ 5. pancreas	f.	reabsorbs water and minerals
_____ 6. esophagus	g.	secretes enzymes that digest carbohydrate, fat, and protein; absorbs nutrients
_____ 7. small intestine	h.	chews food and mixes it with saliva
_____ 8. gallbladder		

Answer these controversy questions:

1. Why is the term *moderation* difficult to define in relationship to alcohol consumption?

2. How does food slow the absorption of alcohol?

3. What advice would you give someone who wants to drink alcohol socially but not become intoxicated?

4. Why are nutrient deficiencies an inevitable consequence of alcohol abuse?

5. Where else can the health-promoting components of red wine be found?

Study Aids

1. Use Table 3-1 on page 84 in your textbook to study the chemical digestion of nutrients.
2. On the following page, label the names of the accessory organs that aid digestion, on the left, and the names of the digestive tract organs that contain the food, on the right.

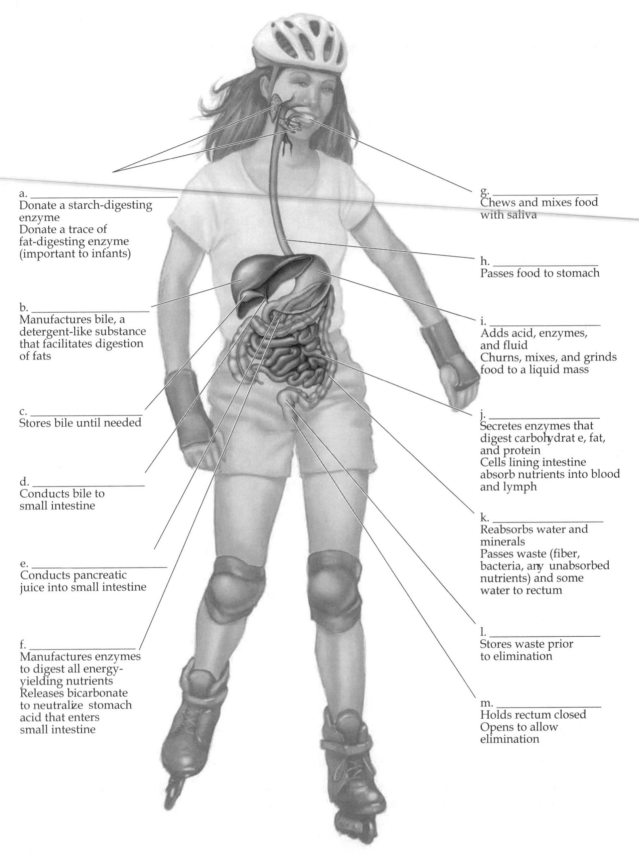

a. _____
Donate a starch-digesting
enzyme
Donate a trace of
fat-digesting enzyme
(important to infants)

b. _____
Manufactures bile, a
detergent-like substance
that facilitates digestion
of fats

c. _____
Stores bile until needed

d. _____
Conducts bile to
small intestine

e. _____
Conducts pancreatic
juice into small intestine

f. _____
Manufactures enzymes
to digest all energy-
yielding nutrients
Releases bicarbonate
to neutralize stomach
acid that enters
small intestine

g. _____
Chews and mixes food
with saliva

h. _____
Passes food to stomach

i. _____
Adds acid, enzymes,
and fluid
Churns, mixes, and grinds
food to a liquid mass

j. _____
Secretes enzymes that
digest carbohydrat e, fat,
and protein
Cells lining intestine
absorb nutrients into blood
and lymph

k. _____
Reabsorbs water and
minerals
Passes waste (fiber,
bacteria, any unabsorbed
nutrients) and some
water to rectum

l. _____
Stores waste prior
to elimination

m. _____
Holds rectum closed
Opens to allow
elimination

Sample Test Items

Comprehension Level Items:

1. Genes direct the making of a piece of protein machinery, which is often a (an) _____, that helps to do the cell's work.

   a. antibody
   b. hormone
   c. enzyme
   d. antigen

2. Every cell continuously draws _____ from the body fluids.

   a. oxygen
   b. nutrients
   c. carbon dioxide
   d. a and b
   e. b and c

3. All blood leaving the digestive system is routed directly to the:

   a. pancreas.
   b. liver.
   c. kidneys.
   d. gallbladder.

4. The blood often serves as an indicator of disorders caused by dietary deficiencies or imbalances of vitamins and minerals.

   a. true
   b. false

5. Hormones act as messengers that stimulate various organs to take appropriate actions.

   a. true
   b. false

6. When the pancreas detects a high concentration of the blood's sugar, glucose, it:

   a. secretes a hormone called glucagon.
   b. stimulates the liver to release glucose into the blood.
   c. stimulates special nerve cells in the hypothalamus.
   d. releases a hormone called insulin.

7. Which of the following statements is **not** true concerning the hormonal system?

   a. It regulates the menstrual cycle in women.
   b. It helps to regulate hunger and appetite.
   c. It is affected by nutrition very little.
   d. It regulates the body's reaction to stress.

8. Which of the following occur(s) as part of the stress response?

   a. The pupils of the eyes widen.
   b. The heart races to rush oxygen to the muscles.
   c. The digestive system speeds up.
   d. a and b
   e. a and c

9. The mechanical job of digestion is to:

    a. chew and crush foods.   d. a and c
    b. add fluid.   e. a and b
    c. secrete enzymes.

10. The major site for digestion of food and absorption of nutrients is the:

    a. small intestine.   c. stomach.
    b. colon.   d. liver.

11. Which of the following ensures that the digestive tract lining will not be digested?

    a. enzymes   c. bile
    b. mucus   d. hydrochloric acid

For questions 12-16, match the digestive organs, listed on the left, with their appropriate functions, listed on the right.

12. _____ small intestine   a. releases bicarbonate to neutralize stomach acid
                                  b. churns, mixes, and grinds food to a liquid mass
13. _____ stomach   c. cells of its lining absorb nutrients into blood and lymph
14. _____ pancreas   d. passes any unabsorbed nutrients, waste (fiber, bacteria), and some
                                    water to rectum
15. _____ large intestine   e. manufactures bile to help digest fats
16. _____ anus   f. holds rectum closed

17. Which of the following is(are) the main task(s) of the colon?

    a. to absorb vitamins   d. a and b
    b. to absorb water   e. b and c
    c. to absorb some minerals

18. The digestive tract can adjust to whatever mixture of foods is presented to it.

    a. true   b. false

19. Characteristics of the cells of the intestinal lining include:

    a. they are very efficient.   d. a and b
    b. they are selective.   e. b and c
    c. they are inefficient.

20. Which of the following nutrients should be consumed in intervals throughout the day based on the body's storage systems?

    a. fat   c. sodium
    b. carbohydrate   d. protein

21. Which of the following cells are the first to defend the body tissues against invaders?

    a. T-cells   c. phagocytes
    b. B-cells   d. lymphocytes

Chapter 4 - The Carbohydrates
Sugar, Starch, Glycogen, and Fiber

Chapter Objectives

After completing this chapter, you should be able to:

1. Distinguish among the various carbohydrates found in foods and in the human body.

2. Differentiate between complex and simple carbohydrates and their effects on the body.

3. Discuss the roles of fiber-rich foods in the maintenance of the body's health and identify foods rich in fiber.

4. Describe the symptoms and management of lactose intolerance.

5. Describe the body's use of glucose to provide energy or to make glycogen and fat.

6. Differentiate between type 1 and type 2 diabetes and strategies for their management.

7. Identify the two different types of hypoglycemia, their symptoms, and treatment.

8. Assess the role sugar and alternative sweeteners play in one's diet and describe their impact on health (Controversy 4).

Key Concepts

✓ Through photosynthesis, plants combine carbon dioxide, water, and the sun's energy to form glucose. Carbohydrates are made of carbon, hydrogen, and oxygen held together by energy-containing bonds: *carbo* means "carbon"; *hydrate* means "water."

✓ Glucose is the most important monosaccharide in the human body. Most other monosaccharides and disaccharides become glucose in the body.

✓ Starch is the storage form of glucose in plants and is also nutritive for human beings.

✓ Glycogen is the storage form of glucose in animals and human beings.

✓ Human digestive enzymes cannot break the bonds in fiber, so most of it passes through the digestive tract unchanged. Some fiber, however, is susceptible to fermentation by bacteria in the colon.

✓ The body tissues use carbohydrates for energy and other functions; the brain and nerve tissues prefer carbohydrate as fuel. Nutrition authorities recommend a diet based on foods rich in complex carbohydrates and fiber.

✓ Fiber-rich diets benefit the body by helping to normalize blood cholesterol and blood glucose and by maintaining healthy bowel function. They are also associated with healthy body weight.

✓ Foods rich in soluble viscous fibers help control blood cholesterol.

✓ Foods rich in viscous fibers help to modulate blood glucose concentrations.

✓ Fibers in foods help to maintain digestive tract health.

✓ Diets that are adequate in fiber assist the eater in maintaining a healthy body weight.

- ✓ Most adults need between 24 and 38 grams of total fiber each day, but few consume this amount. Fiber needs are best met with whole foods. Purified fiber in large doses can have undesirable effects. Fluid intake should increase with fiber intake.

- ✓ With respect to starch and sugars, the main task of the various body systems is to convert them to glucose to fuel the cells' work. Fermentable fibers may release gas as they are broken down by bacteria in the intestine.

- ✓ Lactose intolerance is a common condition in which the body fails to produce sufficient amounts of the enzyme needed to digest the sugar of milk. Uncomfortable symptoms result and can lead to milk avoidance. Lactose-intolerant people and those allergic to milk need milk alternatives that contain the calcium and vitamins of milk.

- ✓ Without glucose, the body is forced to alter its uses of protein and fats. To help supply the brain with glucose, the body breaks down protein to make glucose and converts its fats into ketone bodies, incurring ketosis.

- ✓ Glycogen is the body's storage form of glucose. The liver stores glycogen for use by the whole body. Muscles have their own glycogen stock for their exclusive use. The hormone glucagon acts to liberate stored glucose from liver glycogen.

- ✓ Blood glucose regulation depends mainly on the hormones insulin and glucagon. Most people have no problem regulating their blood glucose when they consume mixed meals at regular intervals.

- ✓ The glycemic index is a measure of blood glucose response to foods relative to the response to a standard food. The glycemic load is the product of the glycemic index multiplied by the carbohydrate content of a food. The concept of good and bad foods based on the glycemic response is an oversimplification.

- ✓ The liver has the ability to convert glucose into fat; under normal conditions, most excess glucose is stored as glycogen or used to meet the body's immediate needs for fuel.

- ✓ Diabetes is an example of the body's abnormal handling of glucose. It is a major threat to health and life, and its prevalence is rapidly increasing. Prediabetes silently threatens health.

- ✓ Type 1 diabetes is an autoimmune disease that attacks the pancreas. Inadequate insulin leaves blood glucose high and cells undersupplied with glucose energy. People with type 1 diabetes depend on external sources of insulin.

- ✓ Type 2 diabetes is a growing problem. The risk of developing it rises with weight gain, aging, and physical inactivity, and falls with a nutritious diet as part of a healthy lifestyle.

- ✓ Diet plays a central role in controlling diabetes and the illnesses that accompany it . A person with diabetes must establish patterns of eating, exercise, and medication to control blood glucose.

- ✓ Postprandial hypoglycemia is an uncommon medical condition in which blood glucose falls too low. It can be a warning of organ damage or serious disease.

Summing Up

Carbohydrates are the first link in the food chain and are obtained almost exclusively from

(1)_____. Through (2)_____, plants combine carbon dioxide, water,

and the sun's energy to form (3)_____, from which the human body can obtain energy.

Carbohydrates are made of carbon, hydrogen, and (4)_____. Six sugar molecules are

important in nutrition, including three (5)_____ and three disaccharides. The monosaccharides include glucose, galactose, and (6)_____; the disaccharides include (7)_____, maltose, and (8)_____.

Glucose may be strung together in long strands of thousands of glucose units to form (9)_____, which include starch, glycogen, and most of the (10)_____. Some of the best known insoluble fibers are (11)_____ and hemicellulose, while soluble fibers include (12)_____ and gums.

Glucose from carbohydrate is the preferred (13)_____ for most body functions; nerve cells, including those of the (14)_____, depend almost exclusively on glucose for their energy. Unfortunately, (15)_____ carbohydrate is often wrongly accused in popular books as being the "fattening" ingredient of foods. The Committee on DRI recommends (16)_____ percent of total calories from carbohydrates.

Many different forms of fiber exist, each with their specific effects on health. According to the American Dietetic Association, most people need (17)_____ grams of total fiber each day. Fiber needs are best met by eating (18)_____ plant foods, which contain a mix of fiber types; (19)_____ fiber in large doses can have undesirable effects.

Glucose is the basic carbohydrate unit that each (20)_____ of the body uses for energy. If the blood delivers more glucose than the cells need, the liver and (21)_____ cells take up the surplus and build the polysaccharide (22)_____. Without sufficient carbohydrate, the body turns to (23)_____ to make glucose. Also, without sufficient carbohydrate, the body cannot use its (24)_____ in the normal way.

Some people have physical conditions which cause abnormal handling of carbohydrates. These conditions include (25)_____ intolerance, hypoglycemia, and (26)_____. The predominant type of diabetes is (27)_____, characterized by (28)_____ resistance of the body's cells. In contrast, the person with type 1 diabetes secretes no (29)_____. The person with diabetes is especially advised to control body (30)_____ because overweight worsens diabetes. Nutrition intervention plays a central role in managing diabetes, with a focus on controlling (31)_____ intake.

The term (32)_____ refers to abnormally low blood glucose. The two types of hypoglycemia include postprandial and (33)_____. People who experience the symptoms of hypoglycemia may benefit from eating regularly timed, balanced (34)_____ and minimizing (35)_____ beverages.

Chapter Glossary

Crossword Puzzle:

Across:

9. hypoglycemia that occurs after 8 to 14 hours of fasting
10. a disease characterized by elevated blood glucose and inadequate or ineffective insulin, which impairs a person's ability to regulate blood glucose normally
11. a hormone secreted by the pancreas that stimulates the liver to release glucose into the blood when blood glucose concentration dips
13. the process by which green plants make carbohydrates from carbon dioxide and water using the green pigment chlorophyll to capture the sun's energy
14. the type of diabetes in which the pancreas produces no or very little insulin

Down:

1. an unusual drop in blood glucose that follows a meal and is accompanied by symptoms such as anxiety, rapid heartbeat, and sweating
2. the action of carbohydrate and fat in providing energy that allows protein to be used for purposes it alone can serve
3. long chains of sugar units arranged to form starch or fiber; also called polysaccharides
4. a condition in which a normal or high amount of insulin produces a less than normal response by the tissues; metabolic consequence of obesity
5. the green pigment of plants that captures energy from sunlight for use in photosynthesis
6. impaired ability to digest lactose due to a lack of the enzyme lactase
7. sugars, including both single sugar units and linked pairs of sugar units
8. a blood glucose concentration below normal, a symptom that may indicate any of several diseases, including impending diabetes
12. a small fat fragment produced by the fermenting action of bacteria on viscous, soluble fibers

Matching Exercise 1:

_____ 1. monosaccharides

_____ 2. maltose

_____ 3. chelating agents

_____ 4. hemorrhoids

_____ 5. starch

_____ 6. appendicitis

_____ 7. polysaccharides

_____ 8. glucose

_____ 9. soluble fibers

_____ 10. diverticula

_____ 11. glycogen

_____ 12. fibers

_____ 13. sucrose

_____ 14. granules

_____ 15. insoluble fibers

a. compounds composed of long strands of glucose units linked together

b. molecules that attract or bind with other molecules and are therefore useful in either preventing or promoting movement of substances from place to place

c. a disaccharide composed of glucose and fructose

d. food components that readily dissolve in water and often impart gummy or gel-like characteristics to foods

e. a disaccharide composed of two glucose units

f. the tough, fibrous structures of fruits, vegetables, and grains; indigestible food components that do not dissolve in water

g. a plant polysaccharide composed of glucose; highly digestible by human beings

h. a polysaccharide composed of glucose, made and stored by liver and muscle tissues of human beings and animals as a storage form of glucose

i. packages of starch molecules; various plant species make these in varying shapes

j. inflammation and/or infection of the appendix, a sac protruding from the intestine

k. a single sugar used in both plant and animal tissues for energy

l. sacs or pouches that balloon out of the intestinal wall, caused by weakening of the muscle layers that encase the intestine

m. the indigestible parts of plant foods, comprised mostly of cellulose, hemicellulose, and pectin

n. swollen, hardened (varicose) veins in the rectum, usually caused by the pressure resulting from constipation

o. single sugar units

_____ 1. constipation

_____ 2. disaccharides

_____ 3. galactose

_____ 4. lactase

_____ 5. fructose

_____ 6. prediabetes

_____ 7. sugars

_____ 8. carbohydrates

_____ 9. lactose

_____ 10. glycemic index

_____ 11. type 2 diabetes

_____ 12. insulin

_____ 13. ketone bodies

_____ 14. ketosis

_____ 15. resistant starch

a. a condition in which blood glucose levels are higher than normal but not high enough to be diagnosed as diabetes

b. a disaccharide composed of glucose and galactose; sometimes known as milk sugar

c. pairs of single sugars linked together

d. a monosaccharide; sometimes known as fruit sugar

e. monosaccharide; part of the disaccharide lactose

f. difficult, incomplete, or infrequent bowel movements; associated with discomfort in passing feces from the body

g. an undesirable high concentration of ketone bodies, such as acetone, in the blood or urine

h. acidic, fat-related compounds that can arise from the incomplete breakdown of fat when carbohydrate is not available

i. hormone secreted by the pancreas in response to a high blood glucose concentration

j. the intestinal enzyme that splits the disaccharide lactose to monosaccharides during digestion

k. the type of diabetes in which the person makes plenty of insulin but the body cells resist insulin's action

l. simple carbohydrates; that is, molecules of either single sugar units or pairs of those sugar units bonded together

m. ranking of foods according to their potential for raising blood glucose relative to a standard such as glucose or white bread

n. compounds composed of single or multiple sugars

o. the fraction of starch in a food that is digested slowly, or not at all, by human enzymes

Exercises

Answer these chapter study questions:

1. From a nutrition perspective, how do fruits differ from concentrated sweets?

2. How does fiber help maintain a healthy body weight?

3. Describe how the body adjusts after a meal when the blood glucose level rises.

4. What happens when the body faces a severe carbohydrate deficit?

5. Why can some people with lactose intolerance consume and tolerate yogurt or aged cheese?

6. Describe insulin resistance that is a characteristic of type 2 diabetes.

Complete these short answer questions:

1. The only animal-derived food that contains significant amounts of carbohydrate is _____.

2. Carbohydrates are made of:

   a.   c.

   b.

3. The monosaccharides include:

   a.   c.

   b.

4. The disaccharides include:

    a.   c.

    b.

5. The polysaccharides include:

    a.   c.

    b.

6. The fibers of a plant contribute the supporting structures of its:

    a.   c.

    b.

7. Soluble fiber is commonly found in:

    a.   d.

    b.   e.

    c.

8. Groups of individuals who are especially vulnerable to the adverse effects of excess fiber include:

    a.

    b.

    c.

9. A balanced diet that is high in _____ carbohydrate helps control body weight and maintain lean tissue.

10. An average-size person needs around _____ grams of carbohydrate per day to insure complete sparing of body protein and avoidance of ketosis.

Solve these problems:

1. Estimate the grams of dietary fiber and carbohydrate supplied by the following menu. Refer to the Food Feature on pages 131-136 for carbohydrate and fiber information.

Breakfast	Fiber (grams)	Carbohydrate (grams)
1 cup skim milk	_____	_____
1 slice whole wheat toast	_____	_____
1 oz. cheddar cheese	_____	_____
1 small orange	_____	_____
Lunch		
Salad with:		
2 cups raw lettuce	_____	_____
1 medium tomato	_____	_____
2 tbsp. oil and vinegar dressing	_____	_____
1 small apple	_____	_____
Dinner		
3 oz. chicken breast	_____	_____
½ cup raw broccoli	_____	_____
1 small baked potato with skin	_____	_____
1 slice light rye bread	_____	_____
1 cup plain yogurt with	_____	_____
½ cup blackberries	_____	_____
Total:	_____	_____

2. Use the ingredient label below to respond to the questions which follow.

> **IN GREDI ENTS:** Bleached flour, sugar, partially hydrogenated vegetable shortening, dextrose, water, corn syrup, carob, whey blend, cornstarch, salt, sodium bicarbonate, lecithin, artificial flavoring and artificial colors.

   a. List all of the sugars which appear in the above product.

   b. Would you consider this product to be high or low in sugar content? Why?

Answer these controversy questions:

1. What is primarily responsible for the steady, upward trend in U.S. sugar consumption?

2. What is the relationship between sugar and the cause of type 2 diabetes?

3. How would you respond to someone who makes the statement that sugary foods cause children to become hyperactive and unruly?

4. If a person consumes 3,000 calories a day, how many calories from added sugars could he/she consume without exceeding the recommendations of many worldwide governing agencies?

5. Why are sugar alcohols safer for teeth than carbohydrate sweeteners?

Study Aids

1. Identify whether the following health effects are characteristic of insoluble or soluble fiber by placing an X in the appropriate column.

		Insoluble	Soluble
a.	slows transit of food through the upper digestive tract	_____	_____
b.	alleviates constipation	_____	_____
c.	slows glucose absorption	_____	_____
d.	lowers blood cholesterol	_____	_____
e.	lowers risk of diverticulosis	_____	_____

2. Use Table 4-2 on page 110 to study the characteristics, food sources, and health effects of soluble and insoluble fiber. Study Table 4-7 on page 127 in your textbook to differentiate between type 1 and type 2 diabetes.

Sample Test Items

Comprehension Level Items:

1. Carbohydrate-rich foods are obtained almost exclusively from animals.

    a. true
    b. false

2. What are carbohydrates made of?

    a. oxygen, nitrogen, and carbon
    b. nitrogen, hydrogen, and carbon
    c. hydrogen, carbon, and oxygen
    d. carbon dioxide, hydrogen, and oxygen

3. Fructose occurs in all of the following **except**:

    a. honey.
    b. table sugar.
    c. fruit.
    d. milk.

4. Maltose consists of:

    a. glucose and galactose.
    b. galactose and fructose.
    c. two glucose units.
    d. lactose and glucose.

5. Which of the following rarely occurs free in nature?

    a. fructose
    b. galactose
    c. sucrose
    d. glucose

6. Complex carbohydrates are called:

    a. natural sugars.
    b. monosaccharides.
    c. polysaccharides.
    d. complex sugars.

7. The storage form of glucose found in plants is called:

   a. starch.
   b. chlorophyll.

   c. fiber.
   d. glycogen.

8. Pectin is commonly used:

   a. to thicken jelly.
   b. to add a pleasing consistency to foods.
   c. to make milk more digestible.

   d. a and b
   e. a and c

For questions 9–11, match the terms on the right with their definitions on the left.

9. _____ food components that readily dissolve in water and often impart gummy or gel-like characteristics to foods

10. _____ the indigestible parts of plant foods, largely non-starch polysaccharides

11. _____ the tough, fibrous structures of fruits, vegetables, and grains

   a. fibers
   b. residue
   c. roughage
   d. insoluble fibers
   e. soluble fibers

12. The human brain depends almost exclusively on _____ for its energy.

   a. fat
   b. carbohydrate

   c. alcohol
   d. protein

13. Fiber promotes maintenance of proper weight through all of the following effects **except**:

   a. absorbing water and creating a feeling of fullness.
   b. donating few calories.
   c. speeding up movement of food through the upper digestive tract.
   d. displacing calorie-dense concentrated fats and sweets from the diet.

14. Which of the following parts of a wheat kernel is rich in both nutrients and fiber?

   a. endosperm
   b. germ

   c. husk
   d. bran

15. The body's form of stored glucose is called:

   a. glycogen.
   b. starch.

   c. glucagon.
   d. ketones.

16. Which of the following would you choose to lower your blood cholesterol level?

   a. purified fiber
   b. wheat bran

   c. brown rice
   d. oat bran

17. Which of the following foods would be most easily digested based on its starch content?

   a. corn
   b. lima beans

   c. white bread
   d. potato

18. On drinking milk, a person with lactose intolerance is likely to experience:

    a. vomiting.
    b. diarrhea.
    c. excessive gas.

    d. a and b
    e. b and c

19. Possible effects of diabetes include:

    a. blindness.
    b. kidney failure.
    c. cancer of the pancreas.

    d. a and b
    e. b and c

20. Carbohydrate-containing foods appear in all of the following food groups **except**:

    a. milk, cheese and yogurt.
    b. vegetables.

    c. grains.
    d. fats and oils.

21. Any one of the following fruit choices contributes about _____ grams of carbohydrate: ½ cup juice, a small banana, or ½ cup canned fruit.

    a. 5
    b. 12

    c. 15
    d. 18

22. Neither honey nor sugar is a good source of any nutrients other than carbohydrates.

    a. true

    b. false

23. The glycemic index may be important to people with:

    a. diabetes.
    b. hypoglycemia.

    c. lactose intolerance.
    d. diverticulosis.

24. The most appropriate diet for managing diabetes is:

    a. controlled in total carbohydrate.
    b. low in saturated and *trans* fat.
    c. high in protein.

    d. a and b
    e. a and c

25. To magnify the sweetness of foods without adding extra sugar and boosting the calories, you would:

    a. serve sweet food cold.
    b. add sweet spices such as cinnamon and nutmeg.
    c. add a tiny pinch of salt.
    d. a and b
    e. b and c

Application Level Items:

26. Someone makes the statement that eating fruit is the same as eating a candy bar. Your response would be:

    a. fruits differ from candy bars in nutrient density.
    b. the sugars of fruits are diluted in large volumes of water.
    c. candy bars are packaged in fiber.
    d. a and b
    e. b and c

27. Which of the following would you recommend as a calcium source for someone with lactose intolerance?

    a. milk
    b. yogurt with added milk solids
    c. aged cheese

    d. a and b
    e. b and c

28. Why would you recommend that someone not add purified fiber to the diet?

    a. It might displace nutrients from the diet.
    b. It might easily be taken to the extreme.
    c. It is very expensive and adversely affects the taste of foods.
    d. a and b
    e. b and c

29. Which of the following types of flour contains all of the nutritive portions of the original grain?

    a. white flour
    b. unbleached flour

    c. whole-grain flour
    d. wheat flour

30. What argument(s) would you use against following a low-carbohydrate diet?

    a. The body uses protein to make energy.
    b. The body uses up its fat as an energy source.
    c. The body cannot use its fat in the normal way.
    d. a and b
    e. a and c

31. Use the Food Feature in this chapter of your textbook to determine the grams of carbohydrate and calories from carbohydrate in the following meal: 1 cup nonfat skim milk; 3 oz. sirloin steak; $1/3$ cup white rice; ½ cup cooked carrots; 1 slice whole-wheat bread; 1 small apple.

    a. 52 grams CHO; 208 calories
    b. 58 grams CHO; 232 calories

    c. 62 grams CHO; 248 calories
    d. 72 grams CHO; 288 calories

32. A person consumes 2,500 calories a day and wants to have 60% of his calories come from carbohydrate. How many grams of carbohydrate should this person consume?

    a. 250
    b. 375

    c. 400
    d. 525

Answers

Summing Up

1. plants
2. photosynthesis
3. glucose
4. oxygen
5. monosaccharides
6. fructose
7. sucrose
8. lactose
9. polysaccharides
10. fibers
11. cellulose
12. pectin
13. fuel
14. brain
15. complex
16. 45-65
17. 20-35
18. whole
19. purified
20. cell
21. muscle
22. glycogen
23. protein
24. fat
25. lactose
26. diabetes
27. type 2
28. insulin
29. insulin
30. fatness
31. carbohydrate
32. hypoglycemia
33. fasting
34. meals
35. alcoholic

Chapter Glossary

Crossword Puzzle:

1. postprandial hypoglycemia
2. protein-sparing action
3. complex carbohydrates
4. insulin resistance
5. chlorophyll
6. lactose intolerance
7. simple carbohydrates
8. hypoglycemia
9. fasting hypoglycemia
10. diabetes
11. glucagon
12. butyrate
13. photosynthesis
14. type 1 diabetes

Matching Exercise 1:

1. o
2. e
3. b
4. n
5. g
6. j
7. a
8. k
9. d
10. l
11. h
12. m
13. c
14. i
15. f

Matching Exercise 2:

1. f
2. c
3. e
4. j
5. d
6. a
7. l
8. n
9. b
10. m
11. k
12. i
13. h
14. g
15. o

Exercises

Chapter Study Questions:

1. The sugars in fruits are diluted in large volumes of water and packaged with fiber, phytochemicals, vitamins and minerals needed by the body. Concentrated sweets are much less nutrient dense and may also be high in fat.

2. Because high-fiber foods absorb water from the digestive juices, swell and create a feeling of fullness; in addition, they donate few calories and displace calorie-dense concentrated fats and sweets from the diet. Some fibers also slow movement of food through the upper digestive tract so that you feel fuller for a longer period of time.

Chapter 5 - The Lipids
Fats, Oils, Phospholipids, and Sterols

Chapter Objectives

After completing this chapter, you should be able to:

1. Explain the ways lipids are useful, both in foods and in the body.

2. Describe the structure of a triglyceride, noting the differences between saturated and unsaturated fats.

3. Identify the important roles that phospholipids and sterols play in the body.

4. Summarize the processes of lipid digestion, absorption, transport, storage, and utilization in the body.

5. Describe current recommendations for lipid intakes and ways to decrease LDL and increase HDL cholesterol.

6. Describe the roles of the essential polyunsaturated fatty acids and current recommendations for omega-3 fatty acid intake.

7. Explain why manufacturers frequently hydrogenate fats and discuss health implications of consuming *trans* fatty acids.

8. List the arguments for and against the use of fat replacers.

9. Plan a diet containing fat in the right kinds and in the recommended amounts to provide optimal health and pleasure in eating.

10. Describe the most recent dietary guidelines for fat and identify characteristics of the Mediterranean diet that support heart health (Controversy 5).

Key Concepts

✓ Lipids not only serve as energy reserves but also cushion the vital organs, protect the body from temperature extremes, carry the fat-soluble nutrients and phytochemicals, serve as raw materials, and provide the major component of which cell membranes are made.

✓ Lipids provide more energy per gram than carbohydrate and protein, enhance the aromas and flavors of foods, and contribute to satiety, or a feeling of fullness, after a meal.

✓ The body combines three fatty acids with one glycerol to make a triglyceride, its storage form of fat. Fatty acids in food influence the composition of fats in the body.

✓ Fatty acids are energy-rich carbon chains that can be saturated (filled with hydrogens) or monounsaturated (with one point of unsaturation) or polyunsaturated (with more than one point of unsaturation). The degree of saturation of the fatty acids in a fat determines the fat's softness or hardness.

✓ Phospholipids, including lecithin, play key roles in cell membranes; sterols play roles as part of bile, vitamin D, the sex hormones, and other important compounds.

✓ In the stomach, fats separate from other food components. In the small intestine, bile emulsifies the fats, enzymes digest them, and the intestinal cells absorb them.

✓ Small lipids travel in the bloodstream unassisted. Large lipids are incorporated into chylomicrons for transport in the lymph and blood. Blood and other body fluids are watery, so fats need special transport vehicles—the lipoproteins—to carry them in these fluids.

✓ When low on fuel, the body draws on its stored fat for energy. Carbohydrate is necessary for the complete breakdown of fat.

✓ The chief lipoproteins are chylomicrons, VLDL, LDL, and HDL. Blood LDL and HDL concentrations are among the major risk factors for heart disease.

✓ Elevated blood cholesterol is a risk factor for cardiovascular disease. Among major dietary factors that raise blood cholesterol, saturated fat and *trans* fat intakes are the most influential. Dietary cholesterol raises blood cholesterol to a lesser degree.

✓ Trimming fat from food trims calories and, often, saturated fat and *trans* fat as well.

✓ Dietary measures to lower LDL in the blood involve reducing saturated fat and *trans* fat and substituting monounsaturated and polyunsaturated fats. Cholesterol-containing foods are nutritious and are best used in moderation by most people.

✓ Two polyunsaturated fatty acids, linoleic acid (an omega-6 fatty acid) and linolenic acid (an omega-3 acid), are essential nutrients used to make substances that perform many important functions. The omega-6 family includes linoleic acid and arachidonic acid. The omega-3 family includes linolenic acid, EPA, and DHA. The principal food source of EPA and DHA is fish, but some species have become contaminated with environmental pollutants.

✓ Vegetable oils become more saturated when they are hydrogenated. Hydrogenated fats resist rancidity better, are firmer textured, and have a higher smoking point than unsaturated oils; but they also lose the health benefits of unsaturated oils.

✓ The process of hydrogenation also creates *trans* fatty acids. *Trans* fats act like saturated fats in the body. Consumers should not lose sight of saturated fats as the main dietary risk factor for heart and artery disease.

✓ Fats added to foods during preparation or at the table are a major source of fat in the diet.

✓ Meats account for a large proportion of the hidden fat and saturated fat in many people's diets. Most people consume meat in larger amounts than recommended.

✓ The choice between whole and fat-free milk products can make a large difference to the fat and saturated fat content of a diet. Cheeses are a major contributor of saturated fat.

✓ Fat in breads and cereals can be well hidden. Consumers must learn which foods of this group contain fats.

Summing Up

The lipids in foods and in the human body fall into (1)_____ classes. About 95 percent of the lipids are (2)_____. Other classes of the lipid family include the phospholipids and the (3)_____.

Although too much fat in the diet has the potential to be (4)_____, lipids serve many valuable functions in the body. Fat is the body's chief storage form for (5)_____ from food eaten in excess of need. In addition, fat surrounds and cushions all the body's vital

(6)_____ and insulates the body from extremes in (7)_____. Some essential

nutrients, such as the essential (8)_____ and the fat-soluble vitamins A, D, E, and

(9)_____, are found primarily in foods that contain fat.

Fatty acids are the major constituent of (10)_____, which are the chief form of fat, and

they differ from one another in two ways: in (11)_____ length and in degree of

(12)_____. The more unsaturated a fat, the more (13)_____ it is at room

temperature. Generally, vegetable and (14)_____ oils are rich in polyunsaturates.

However, certain vegetable oils, including coconut and (15)_____, are highly saturated.

Within the body, many fats travel from place to place in blood as passengers in

(16)_____. A lipoprotein's density changes with its lipid and (17)_____

contents. Two types of lipoproteins of concern to health care providers include (18)_____

and (19)_____. Elevated HDL concentrations are associated with a

(20)_____ risk of heart attack. High blood (21)_____ is an indicator of risk

for CVD and the main dietary factors associated with elevated blood cholesterol are high

(22)_____ fat and *trans* fat intakes.

Two fatty acids, including linoleic and (23)_____, must be supplied by the diet and

are, therefore, (24)_____ fatty acids. Points of unsaturation in fatty acids are vulnerable to

attack by (25)_____ and when the unsaturated points are oxidized, the oils become

(26)_____. Cooking oils should be stored in tightly covered containers that exclude

(27)_____. One way to prevent spoilage of unsaturated fats and also to make them harder

is to change their fatty acids chemically by (28)_____. The process of

hydrogenation creates (29)_____, which pose a risk to the health of

the heart and arteries because they act like saturated fats in the body. The DRI Committee

recommendation is to consume as little *trans* fat as possible.

Fats added to foods during (30)_____ or at the table are a major source of fat in the

diet. Fats may be (31)_____ on foods, such as the fat trimmed from a steak, or they may be

invisible, such as the fat in biscuits or olives. (32)_____ fats are on the rise in U.S. diets.

Chapter Glossary

Matching Exercise:

_____ 1. phospholipids

_____ 2. fats

_____ 3. lipid

_____ 4. oils

_____ 5. cholesterol

_____ 6. emulsifier

_____ 7. sterols

_____ 8. polyunsaturated fatty acid

_____ 9. bile

_____ 10. saturated fatty acid

_____ 11. monounsaturated fatty acid

_____ 12. lecithin

_____ 13. unsaturated fatty acid

_____ 14. monoglycerides

_____ 15. glycerol

_____ 16. triglycerides

_____ 17. cardiovascular disease

_____ 18. saturated fats

a. lipids that are solid at room temperature (70° F or 25° C)

b. an emulsifier made by the liver from cholesterol and stored in the gallbladder

c. one of the three main classes of dietary lipids and the chief form of fat in foods and in the human body

d. a fatty acid containing one point of unsaturation

e. a substance that mixes with both fat and water and permanently disperses the fat in the water, forming an emulsion

f. one of the three main classes of dietary lipids; these lipids are similar to triglycerides, but each has a phosphorus-containing acid in place of one of the fatty acids

g. lipids that are liquid at room temperature (70° F or 25° C)

h. a fatty acid carrying the maximum possible number of hydrogen atoms

i. products of the digestion of lipids

j. a phospholipid manufactured by the liver and also found in many foods; a major constituent of cell membranes

k. a family of organic compounds soluble in organic solvents but not in water

l. a fatty acid that lacks some hydrogen atoms and has one or more points of unsaturation

m. a fatty acid with two or more points of unsaturation

n. an organic compound, three carbons long, of interest here because it serves as the backbone for triglycerides

o. one of the three main classes of dietary lipids, with a structure similar to that of cholesterol

p. a member of the group of lipids known as sterols; a soft waxy substance made in the body for a variety of purposes and also found in animal-derived foods

q. triglycerides in which most of the fatty acids are saturated

r. disease of the heart and blood vessels

Crossword Puzzle:

Word Bank:

chylomicrons	hydrogenation	oxidation
dietary antioxidant	LDL	satiety
essential fatty acids	lipoproteins	smoking point
fatty acids	omega-six fatty acid	*trans* fatty acids
HDL	omega-three fatty acid	

	Across:		Down:
1.	a substance in food that significantly decreases the damaging effects of reactive compounds, such as reactive forms of oxygen and nitrogen, on tissue functioning	2.	interaction of a compound with oxygen
		3.	a polyunsaturated fatty acid with its endmost double bond six carbons from the end of the carbon chain
5.	a polyunsaturated fatty acid with its endmost double bond three carbons from the end of its carbon chain	4.	clusters formed when lipids from a meal are combined with carrier proteins in the intestinal lining
9.	clusters of lipids associated with protein, which serve as transport vehicles for lipids in blood and lymph	6.	lipoproteins containing a large proportion of protein that return cholesterol from tissues to the liver for dismantling and disposal
11.	lipoproteins containing a large proportion of cholesterol that transport lipids from the liver to other tissues such as muscle and fat	7.	organic acids composed of carbon chains of various lengths
12.	the feeling of fullness or satisfaction that people experience after meals	8.	fatty acids with unusual shapes that can arise when polyunsaturated oils are hydrogenated
13.	the process of adding hydrogen to unsaturated fatty acids to make fat more solid and resistant to the chemical change of oxidation	10.	the temperature at which fat gives off an acrid blue gas
14.	fatty acids that the body needs but cannot make in amounts sufficient to meet physiological needs		

Exercises

Answer these chapter study questions:

1. Why is fat, rather than the carbohydrate glucose, the body's major form of stored energy?

2. Why are oils vulnerable to rancidity and how can rancidity be retarded?

3. Why do vegetable oils make up most of the added fat in the U.S. diet?

4. Identify the best way to raise HDL and dietary measures to decrease LDL.

5. What is the best way to ensure the right balance between omega-3 and omega-6 intakes?

6. Why are elevated HDL concentrations associated with a low risk of heart disease?

Complete these short answer questions:

1. The three major classes of lipids include:

   a.   c.

   b.

2. Triglycerides are made up of three _____ units and _____ glycerol unit.

3. Identify three useful functions of fats in foods.

   a.

   b.

   c.

4. Fatty acids may differ from one another in two ways:

   a.   b.

5. The two fatty acids essential for human beings include:

   a.   b.

6. Identify two vegetable oils which are composed primarily of saturated fats.

    a.   b.

7. Three advantages of hydrogenated oils are:

    a.

    b.

    c.

8. The two major types of lipoproteins of concern to health care professionals include:

    a.   b.

9. The main dietary factors associated with elevated blood cholesterol are high _____ and *trans* fat intakes.

10. These two vegetable oils are particularly rich in monounsaturates:

    a.   b.

Solve these problems:

1. a. Identify the primary type of fatty acid found in each food in the lunch menu below.

	Type of Fat
Fried corn (nacho) chips	
Fresh guacamole	
Cheddar cheese sandwich	
Sesame seeds (from bread)	
Walnut & dried fruit trail mix	
Low-fat milk	

    b. Which of these foods contain primarily healthful fats, and which contain primarily harmful fats?

2. Circle all of the words in the label below which alert you to the fat content of the product.

> **IN GREDI ENTS:** Chicken stock, chicken, wheat flour, corn starch, cream, vegetable oil, salt, chicken fat, water, margarine, monosodium glutamate, soy protein isolate, yeast extract and natural flavoring.

3. Modify the following recipe for Beef Stroganoff by making substitutions for the high fat ingredients.

Traditional Recipe	Modifications:
1 pound ground beef	
½ cup chopped onion	
1 small clove garlic, minced	
1 jar sliced mushrooms, drained	
2 tablespoons butter	
2 tablespoons all-purpose flour	
8 ounce carton sour cream	
¼ cup chili sauce	
½ teaspoon Worcestershire sauce	
1 teaspoon salt	
¼ teaspoon pepper	

Answer these controversy questions:

1. What are some of the problems that may accompany low-fat diets?

2. What are five characteristics of the traditional Mediterranean diet thought to be responsible for its health-promoting characteristics?

3. Explain why nuts may lower heart disease risk.

4. Through what specific mechanisms does olive oil confer protection against heart disease?

5. What would you look for on the label when choosing a margarine most likely to support a healthy heart? Why?

Study Aids

Use Table 5-1 on page 146 in your textbook to study the functions of fat in food and in the human body. Study Figure 5-5 on page 151 of your textbook to recognize the fatty acid composition of common food fats.

Sample Test Items

Comprehension Level Items:

1. Which of the following is **not** one of the three main classes of lipids?

   a. triglycerides
   b. lecithin

   c. phospholipids
   d. sterols

2. Lecithin is an example of a:

   a. phospholipid.
   b. triglyceride.

   c. sterol.
   d. fatty acid.

3. Which of the following is **not** a characteristic of fat in food or in the body?

   a. provides energy reserves
   b. protects the body from temperature extremes
   c. helps cushion vital body organs
   d. decreases the flavor of foods
   e. forms the major material of cell membranes

4. A saturated fatty acid is a fatty acid:

   a. possessing an "empty spot."
   b. having two or more points of unsaturation.
   c. carrying the maximum possible number of hydrogen atoms.
   d. a and b
   e. b and c

5. The essential fatty acids include:

   a. linoleic acid.
   b. linolenic acid.
   c. arachidonic acid.

   d. a and b
   e. b and c

6. Which of the following helps prevent spoilage of oils containing unsaturated fatty acids?

   a. Change them chemically by hydrogenation.
   b. Make them more unsaturated.
   c. Add an antioxidant.
   d. a and b
   e. a and c

7. Which of the following cuts of meat is lowest in fat content?

   a. ground chicken
   b. ground turkey
   c. ground chuck
   d. ground round
   e. ground beef

8. A person who eats a diet too high in saturated fat may incur a greater-than-average risk of developing:

   a. heart disease.
   b. osteoporosis.
   c. cancer.
   d. a and b
   e. a and c

9. Within the body, fats usually travel from place to place mixed with _____ particles.

   a. carbohydrate
   b. protein
   c. glycerol
   d. glycogen

10. Which of the following statements concerning cholesterol is **not** true?

    a. It is the major part of the plaques that narrow the arteries.
    b. It is widespread in the body and necessary to its function.
    c. It is an important sterol in the structure of cell membranes.
    d. It cannot be made by the body and must be consumed in the diet.

11. The main dietary factor associated with elevated blood cholesterol levels is:

    a. high total fat intake.
    b. monounsaturated fats.
    c. dietary cholesterol.
    d. high saturated and *trans* fat intake.

12. Which of the following is the single greatest contributor of saturated fat in the diet?

    a. milk
    b. meat
    c. cheese
    d. butter

13. One teaspoon of oil or shortening donates _____ calories.

    a. 25
    b. 30
    c. 40
    d. 45

14. Which of the following is a true statement regarding olestra?

    a. It has not been approved by the FDA.
    b. It passes through the digestive tract unabsorbed.
    c. It has no side effects.
    d. It has been proven safe for use with growing children.

15. A blood lipid profile reveals much useful information regarding a person's risk of:

    a. diabetes.
    b. cancer.
    c. heart disease.
    d. hypoglycemia.

16. In most people, saturated and *trans* fats raise blood cholesterol _____ dietary cholesterol.

    a. more than
    b. less than
    c. the same as
    d. much less than

17. Saturation refers to the number of _____ the fatty acid chain is holding.

    a. oxygens
    b. hydrogens
    c. glycerols
    d. nitrogens

18. *Trans* fatty acids act like _____ fats in the body.

    a. saturated
    b. unsaturated
    c. monounsaturated
    d. polyunsaturated

19. Linolenic acid, EPA and DHA are examples of:

    a. omega-6 fatty acids.
    b. linoleic acids.
    c. monounsaturated fatty acids.
    d. omega-3 fatty acids.

20. Of the following fats, which is the hardest?

    a. safflower oil
    b. chicken fat
    c. beef tallow
    d. coconut oil

21. All plant oils are less saturated than animal fats.

    a. true
    b. false

22. Most fat digestion takes place in the:

    a. mouth.
    b. stomach.
    c. small intestine.
    d. colon.

23. Which of the following is critical in the process of carrying cholesterol from body cells to the liver for disposal?

    a. HDL
    b. VLDL
    c. LDL
    d. chylomicrons

24. All of the following are major risk factors for CVD **except**

    a. family history.
    b. increasing age.
    c. physical inactivity.
    d. female gender.

25. Physical activity promotes heart health because:

    a. it raises blood LDL concentration.
    b. it raises blood HDL concentration.
    c. a smaller volume of blood is pumped with each heartbeat, reducing the heart's workload.
    d. it decreases circulation.

26. A diet that provides _____% of its calories from fat is currently recommended by health authorities.

    a. 10-25   c. 25-40

    b. 20-35   d. 30-45

Application Level Items:

27. Someone asks you whether he should reduce dietary cholesterol in an effort to prevent cardiovascular disease. An appropriate response would be:

    a. cholesterol, in general, doesn't matter.
    b. it doesn't matter as much as saturated and *trans* fat intake.
    c. dietary cholesterol makes a major contribution.
    d. very few foods contain cholesterol anyway.

28. You are trying to get a group of people to meet the current dietary recommendations for fats. You tell them that one of the most effective steps they can take at home is to:

    a. replace butter with margarine.
    b. eat fewer foods.
    c. eliminate solid, saturated fats used as seasonings.
    d. consume fewer snacks throughout the day.

29. Research suggests that North Americans should consume more omega-3 fatty acids. In order to do this, you would consume more:

    a. beef.   c. corn oil.

    b. safflower oil.   d. salmon.

30. You are trying to convince a friend not to buy fish oil supplements. You would use all of the following as arguments **except**:

    a. concentrated supplements make it easy to overdose on toxic vitamins.
    b. the supplements are expensive.
    c. overdoses may cause heart disease and cancer.
    d. supplements may have toxic concentrations of pesticides.

31. A man's father and grandfather both died of heart disease. After ordering a blood lipid profile, the man's physician told him his LDL was too high and his HDL too low. He sometimes walks for exercise, but he smokes cigarettes daily. Would this person be considered at risk for heart disease?

    a. yes   b. no

32. Which of the following would provide the least amount of *trans* fatty acids?

    a. peanut butter   c. liquid margarine

    b. commercial fried fish   d. potato chips

Answers

Summing Up

1. three
2. triglycerides
3. sterols
4. harmful
5. energy
6. organs
7. temperature
8. fatty acids
9. K
10. triglycerides
11. chain
12. saturation
13. liquid
14. fish
15. palm
16. lipoproteins
17. protein
18. LDL
19. HDL
20. low
21. cholesterol
22. saturated
23. linolenic
24. essential
25. oxygen
26. rancid
27. air
28. hydrogenation
29. *trans* fatty acids
30. preparation
31. visible
32. invisible

Chapter Glossary

Matching Exercise:

1. f
2. a
3. k
4. g
5. p
6. e
7. o
8. m
9. b
10. h
11. d
12. j
13. l
14. i
15. n
16. c
17. r
18. q

Crossword Puzzle:

1. dietary antioxidant
2. oxidation
3. omega-6 fatty acid
4. chylomicrons
5. omega-3 fatty acid
6. HDL
7. fatty acids
8. *trans* fatty acids
9. lipoproteins
10. smoking point
11. LDL
12. satiety
13. hydrogenation
14. essential fatty acids

Exercises

Chapter Study Questions:

1. Glycogen, the stored form of carbohydrate, holds a great deal of water and is quite bulky. Therefore, the body cannot store enough glycogen to provide energy for very long. In contrast, fats pack tightly together without water and can store many more calories of energy in a small space.

2. Oils contain points of unsaturation in the fatty acids and these are vulnerable to attack by oxygen. When the unsaturated points are oxidized, the oils become rancid. To prevent rancidity, oils should be stored in tightly covered containers that exclude air and they should be placed in the refrigerator if stored for long periods of time.

3. Because fast food chains use them for frying, manufacturers add them to processed foods, and consumers tend to choose margarine over butter.

4. To raise HDL, people should increase physical activity; to decrease LDL, people should choose a diet that provides 20-35% of its calories from fat, replace saturated fat with monounsaturated or polyunsaturated fats, and keep saturated and *trans* fat as low as possible.

5. Eat 2 fatty fish meals per week and limit the amount of foods high in omega-6 fatty acids, such as vegetable oils and margarine.

6. Because of the task they perform in scavenging excess cholesterol and phospholipids from the body's tissues for disposal.

Short Answer Questions:

1. (a) triglycerides; (b) phospholipids; (c) sterols

2. fatty acid; one

3. (a) provide a concentrated energy source in foods; (b) enhance food's aroma and flavor; (c) contribute to satiety. Refer to Table 5-1 for additional functions of fats in food.

4. (a) in chain length; (b) in degree of saturation

5. (a) linoleic; (b) linolenic

6. (a) coconut oil; (b) palm oil

7. (a) resist rancidity; (b) firm texture; (c) have a high smoking point, making it suitable for frying

8. (a) LDL; (b) HDL

9. saturated

10. (a) olive; (b) canola

Problem-Solving:

1. a.

	Type of Fat
Fried corn (nacho) chips	*trans*
Fresh guacamole	monounsaturated
Cheddar cheese sandwich	saturated
Sesame seeds (from bread)	polyunsaturated (omega-6)
Walnut & dried fruit trail mix	polyunsaturated (omega-3)
Low-fat milk	saturated

   b. The guacamole (made from avocados), sesame seeds, and walnuts contain healthful fats; the corn chips, cheese, and milk contain harmful fats.

2. Chicken stock; chicken; cream; vegetable oil; chicken fat; margarine

3. Use a lean ground round and grain mixture instead of ground beef; use reduced-calorie margarine in the place of butter; use yogurt instead of sour cream.

Controversy Questions:

1. They are difficult to maintain over time; they are not necessarily low-calorie diets and many people with heart disease need to reduce body fatness; may exclude nutritious foods that provide essential fatty acids, phytochemicals and other nutrients, such as fatty fish, nuts, seeds and vegetable oils.

2. The diet is low in saturated fat, very low in *trans* fat, rich in unsaturated fat, rich in starch and fiber, and rich in nutrients and phytochemicals that support good health.

3. Nuts are low in saturated fats, high in fiber, vegetable protein and the antioxidant vitamin E, and they are high in phytochemicals that act as antioxidants.

4. Olive oil may lower total and LDL cholesterol and not lower or raise HDL; it contains phytochemicals that act as antioxidants; it also lowers LDL cholesterol's vulnerability to oxidation, reduces blood clotting factors, and lowers blood pressure.

5. Look for oil as the first ingredient on the label because that means it is probably low in saturated fat and *trans* fatty acids.

Sample Test Items

1. b (p. 146)	9. b (p. 154)	17. b (p. 149)	25. b (p. 161)
2. a (p. 146)	10. d (p. 152)	18. a (p. 167)	26. b (p. 156)
3. d (pp. 146-148)	11. d (p. 158)	19. d (p. 162)	27. b (p. 158)
4. c (p. 149)	12. c (p. 172)	20. c (p. 149)	28. c (p. 175)
5. d (p. 161)	13. d (p. 170)	21. b (p. 150)	29. d (p. 164)
6. e (pp. 166-167)	14. b (p. 170)	22. c (pp. 152-153)	30. c (pp. 164-165)
7. d (pp. 171-172)	15. c (p. 158)	23. a (p. 157)	31. a (p. 158)
8. e (pp. 155-156)	16. a (p. 158)	24. d (p. 158)	32. c (pp. 166-167)

Chapter 6 - The Proteins and Amino Acids

Chapter Objectives

After completing this chapter, you should be able to:

1. Describe the structure of proteins and explain why adequate amounts of all the essential amino acids are required for protein synthesis.

2. Summarize the processes of protein digestion and absorption in the body.

3. Explain the roles of protein in the body.

4. Describe the pros and cons of protein and amino acid supplements.

5. Discuss the importance of protein quality and describe the concept of mutual supplementation.

6. Calculate your individual DRI for protein and describe how nitrogen balance studies are used in determining protein recommendations.

7. Describe the consequences of both protein deficiency and protein excess.

8. Plan a diet that includes enough, but not too much protein.

9. Compare the positive health aspects of a vegetarian diet with those of a diet that includes meat and describe ways each diet can include adequate nutrients (Controversy 6).

Key Concepts

✓ Proteins are unique among the energy nutrients in that they possess nitrogen-containing amine groups and are composed of 20 different amino acid units. Of the 20 amino acids, some are essential, and some are essential only in special circumstances.

✓ Amino acids link into long strands that coil and fold to make a wide variety of different proteins.

✓ Each type of protein has a distinctive sequence of amino acids and so has great specificity. Often, cells specialize in synthesizing particular types of proteins in addition to the proteins necessary to all cells. Nutrients can greatly affect genetic expression.

✓ Proteins can be denatured by heat, acids, bases, alcohol, or the salts of heavy metals. Denaturation begins the process of digesting food protein and can also destroy body proteins.

✓ Digestion of protein involves denaturation by stomach acid, then enzymatic digestion in the stomach and small intestine to amino acids, dipeptides, and tripeptides.

✓ The cells of the small intestine complete digestion, absorb amino acids and some larger peptides, and release them into the bloodstream for use by the body's cells.

✓ The body needs dietary amino acids to grow new cells and to replace worn-out ones.

✓ The body makes enzymes, hormones, and chemical messengers of the nervous system from its amino acids.

✓ Antibodies are proteins that defend against foreign proteins and other foreign substances within the body.

- ✓ Proteins help to regulate the body's electrolytes and fluids.

- ✓ Proteins buffer the blood against excess acidity or alkalinity.

- ✓ Proteins that clot the blood prevent death from uncontrolled bleeding.

- ✓ When insufficient carbohydrate and fat are consumed to meet the body's energy need, food protein and body protein are sacrificed to supply energy. The nitrogen part is removed from each amino acid, and the resulting fragment is oxidized for energy. No storage form of amino acids exists in the body.

- ✓ Amino acids can be metabolized to protein, nitrogen plus energy, glucose, or fat. They will be metabolized to protein only if sufficient energy is present from other sources. The diet should supply all essential amino acids and a full measure of protein according to guidelines.

- ✓ The body's use of a protein depends in part on the user's health, the protein quality, and the other nutrients and energy taken with it. Digestibility of protein varies from food to food and cooking can improve or impair it.

- ✓ A protein's amino acid assortment greatly influences its usefulness to the body. Proteins lacking essential amino acids can be used only if those amino acids are present from other sources.

- ✓ The amount of protein needed daily depends on size and stage of growth. The DRI recommended intake for adults is 0.8 grams of protein per kilogram of body weight.

- ✓ Protein-deficiency symptoms are always observed when either protein or energy is deficient. Extreme food energy deficiency is marasmus; extreme protein deficiency is kwashiorkor. The two diseases overlap most of the time, and together are called PEM. PEM is not unknown in the United States, where millions live on the edge of hunger.

- ✓ Health risks may follow the overconsumption of protein-rich foods.

Summing Up

Protein is a compound composed of carbon, hydrogen, oxygen, and (1)_____ atoms. Protein is made up of building blocks called (2)_____. About (3)_____ amino acids, each with its different side chain, make up most of the proteins of living tissue. The side chains make the amino acids differ in size, (4)_____, and electrical charge.

There are some amino acids that the healthy adult body makes too slowly or cannot make at all and these are called the (5)_____ amino acids. In the first step of making a protein, each amino acid is hooked to the next. A chemical bond, called a (6)_____ bond, is formed between the amine group end of one and the (7)_____ group end of the next. The (8)_____ of proteins enable them to perform different tasks in the body. Among the most fascinating of the proteins are the (9)_____, which act on other substances to change them chemically .

Proteins can be denatured by heat, radiation, alcohol, (10)_____, bases, or the salts of heavy metals. During digestion of proteins in the stomach, the (11)_____ helps to uncoil the protein's strands so that the stomach enzymes can attack the peptide bonds. The stomach lining, which is made partly of (12)_____, is protected by a coat of (13)_____.

84

Continual digestion, and finally absorption, of protein takes place in the (14)_____. The cells of the small intestine possess separate sites for absorbing different types of (15)_____.

Protein serves many roles and the body needs dietary amino acids to grow new (16)_____ and to maintain fluid and electrolyte balance and (17)_____ balance. Under conditions of inadequate energy or (18)_____, protein will be sacrificed to provide needed (19)_____. When amino acids are degraded for energy, their (20)_____ groups are stripped off and used elsewhere or are incorporated by the liver into (21)_____ to be excreted in the urine.

Generally, amino acids from (22)_____ proteins are best absorbed. If the diet fails to provide enough of an (23)_____ amino acid, the cells begin to adjust their activities almost immediately. A diet that is short in any of the essential amino acids limits (24)_____ synthesis. Consuming the essential amino acids is not a problem for people who eat proteins containing ample amounts of all the essential amino acids, such as those of (25)_____, fish, poultry, cheese, eggs, milk and most (26)_____ products. Another option is to eat a combination of foods from (27)_____ so that amino acids that are low in some foods will be supplied by others—a concept called (28)_____.

The DRI for protein depends on (29)_____ and it is higher for infants and (30)_____ to cover needs for building new tissue. For healthy adults, the DRI recommended intake for protein has been set at (31)_____ grams for each kilogram of body weight.

The combination of protein deficiency and energy deficiency is called (32)_____, which is the most widespread form of malnutrition in the world today. PEM takes two different forms, including kwashiorkor and (33)_____. Overconsumption of protein also poses health risks for the (34)_____, for weakened kidneys, and for the bones.

Two groups of foods in the USDA Food Guide contribute high-quality protein, including the (35)_____ group and the meat group. In addition, the vegetable and (36)_____ groups contribute smaller amounts of proteins. Protein-rich foods carry with them a characteristic array of vitamins and (37)_____, including iron, but they lack others, such as vitamin C and (38)_____.

Chapter Glossary

Matching Exercise:

_____ 1. immunity

_____ 2. peptide bond

_____ 3. polypeptides

_____ 4. conditionally essential amino acid

_____ 5. bases

_____ 6. edema

_____ 7. dipeptides

_____ 8. hormones

_____ 9. enzymes

_____ 10. amino acid pools

_____ 11. acid-base balance

_____ 12. fluid and electrolyte balance

_____ 13. protein turnover

_____ 14. acids

_____ 15. hemoglobin

_____ 16. protein-energy malnutrition

_____ 17. denaturation

_____ 18. antibodies

a. chemical messengers secreted by a number of body organs in response to conditions that require regulation

b. the continuous breakdown and synthesis of body proteins involving recycling of amino acids

c. protein fragments of many (more than ten) amino acids bonded together

d. large proteins of the blood, produced by the immune system in response to invasion of the body by foreign substances (antigens)

e. protection from or resistance to a disease or infection by development of antibodies and by the actions of cells and tissues in response to a threat

f. amino acids dissolved in the body's fluids that provide cells with ready raw materials from which to build new proteins or other molecules

g. the world's most widespread malnutrition problem, including both marasmus and kwashiorkor and states in which they overlap

h. swelling of body tissue caused by leakage of fluid from the blood vessels

i. distribution of fluid and dissolved particles among body compartments

j. an amino acid that is normally nonessential, but must be supplied by the diet in special circumstances when the need for it exceeds the body's ability to produce it

k. the globular protein of red blood cells whose iron atoms carry oxygen around the body

l. a bond that connects one amino acid with another, forming a link in a protein chain

m. compounds that release hydrogens in a watery solution

n. protein catalysts

o. the change in a protein's shape brought about by heat, acids, bases, alcohol, salts of heavy metals, or other agents

p. equilibrium between acid and base concentrations in the body fluids

q. protein fragments that are two amino acids long

r. compounds that accept hydrogens from solutions

Crossword Puzzle:

acidosis	dysentery	mutual supplementation
alkalosis	kwashiorkor	serotonin
amino acid	legumes	textured vegetable protein
buffers	limiting amino acid	tofu
collagen	marasmus	urea
complementary proteins		

Across:	Down:
2. blood alkalinity above normal	1. a curd made from soybean that is rich in protein, often rich in calcium and variable in fat content
6. the calorie-deficiency disease; starvation	
8. compounds that help keep a solution's acidity or alkalinity constant	
9. blood acidity above normal, indicating excess acid	3. plants of the bean, pea, and lentil family that have roots with nodules containing special bacteria
10. the principal nitrogen-excretion product of protein metabolism, generated mostly by removal of amine groups from unneeded amino acids or from amino acids being sacrificed to a need for energy	4. processed soybean protein used in products formulated to look and taste like meat, fish, or poultry
	5. two or more proteins whose amino acid assortments complement each other in such a way that the essential amino acids missing from one are supplied by the other
11. a type of body protein from which connective tissues such as scars, tendons, ligaments, and the foundations of bones and teeth are made	
12. an infection of the digestive tract that causes diarrhea	7. the strategy of combining two incomplete protein sources so that the amino acids in one food make up for those lacking in the other food
13. an essential amino acid present in dietary protein in an insufficient amount, so that it limits the body's ability to build protein	
14. a compound related in structure to (and made from) the amino acid tryptophan	9. one of the building blocks of protein
15. a disease related to protein malnutrition, with a set of recognizable symptoms, such as edema	

Exercises

Answer these chapter study questions:

1. Why is milk given as a first-aid remedy when someone swallows a heavy-metal poison?

2. Describe the conditions under which amino acids would be wasted (i.e., not used to build protein or other nitrogen-containing compounds).

3. Describe the main difference between the dietary inadequacies responsible for marasmus versus kwashiorkor.

4. What are legumes and why are they especially protein-rich food sources?

5. Why is it nutritionally advantageous to cook an egg?

6. Why does the DRI Committee recommend that the diet contain no more than 35% of calories from protein?

Complete these short answer questions:

1. Proteins contain the following types of atoms:

    a.   c.

    b.   d.

2. The nine essential amino acids for adults include:

    a.   f.

    b.   g.

    c.   h.

    d.   i.

    e.

3. Amino acids in a cell can be:

    a.

    b.

    c.

    d.

    e.

4. Protein can undergo denaturation by:

    a.   d.

    b.   e.

    c.   f.

5. Examples of proteins that contain ample amounts of all the essential amino acids include:

    a.   e.

    b.   f.

    c.   g.

    d.

6. Examples of individuals in positive nitrogen balance include:

    a.   b.

7. List the steps necessary to calculate a person's DRI for protein.
    a.
    b.
    c.

8. The food groups in the USDA Food Guide which contribute a significant amount of protein to the diet include:

a.   c.

b.   d.

9. Protein-rich foods are notoriously lacking in the nutrients _____ and _____.

10. Animal protein food sources are high in vitamin _____ and the mineral _____.

Solve these problems:

1. What is the DRI for protein for a 37-year-old female who is 5′4″ tall and weighs 110 pounds?

2. Calculate the number of grams of protein provided in the following menu, using Figure 6-16 on page 211 of your textbook. How does this amount of protein compare with the DRI for protein for the female identified in problem #1?

	Menu Item	Grams of Protein
Breakfast	1 cup skim milk	_____
	1 boiled egg	_____
	1 bagel	_____
Lunch	2 oz. tuna fish	_____
	2 slices whole-wheat bread	_____
	2 tsp. mayonnaise	_____
	½ cup strawberries	_____
Dinner	2 oz. chicken breast	_____
	½ cup broccoli	_____
	1 slice whole-grain bread	_____
	1 cup skim milk	_____
	Total:	_____

Answer these controversy questions:

1. What are some of the different reasons why individuals are vegetarians?

2. What are four conditions that a vegetarian diet may offer protection against?

3. Other than food choices, how do vegetarians differ from non-vegetarians?

4. Identify the dietary constituents that, when consumed in moderate to high amounts, are correlated with colon cancer.

5. Which nutrients require special attention for strict vegetarians?

Study Aids

1. Listed below are some of the top contributors of protein to the U.S. diet. Rank order the foods in the list, with 1 being the greatest protein contributor and 7 being the smallest contributor.

	Food	Rank Order
a.	cheese	_____
b.	milk	_____
c.	eggs	_____
d.	beef	_____
e.	fish	_____
f.	dried beans	_____
g.	poultry	_____

2. Identify whether each of these individuals is in positive nitrogen balance, negative nitrogen balance, or nitrogen equilibrium by placing an X in the appropriate column.

	Positive Balance	Negative Balance	Equilibrium
a. Astronaut	_____	_____	_____
b. Growing child	_____	_____	_____
c. Healthy adult	_____	_____	_____
d. Pregnant woman	_____	_____	_____
e. Surgery patient	_____	_____	_____

3. Use Table 6-1 on page 199 of your textbook to study the functions of protein in the body. Study Figure 6-12 on page 199 of your textbook to compare and contrast the three different energy sources: carbohydrate, protein, and fat.

Sample Test Items

Comprehension Level Items:

1. A strand of amino acids that makes up a protein may contain _____ different kinds of amino acids.

   a. 5
   b. 10
   c. 15
   d. 20

2. Which of the following makes amino acids differ in size, shape, and electrical charge?

   a. the side chains
   b. the amine groups
   c. the acid groups
   d. a and b
   e. b and c

3. Which of the following is **not** considered to be an essential amino acid?

   a. threonine
   b. alanine
   c. lysine
   d. phenylalanine

4. The body does not make a specialized storage form of protein as it does for carbohydrate and fat.

   a. true
   b. false

5. Appropriate roles of protein in the body include all of the following **except**

   a. providing the body's needed energy.
   b. participating in growth and development.
   c. maintaining fluid and electrolyte balance.
   d. maintaining acid-base balance.

6. When there is a surplus of amino acids, the body does all of the following **except**:

   a. removes and excretes their amine groups.
   b. uses the residues to build extra muscle tissue.
   c. uses the residues for immediate energy.
   d. uses the residues to make glucose for storage as glycogen.
   e. uses the residues to make fat for energy storage.

7. Athletes need slightly more protein than the DRI, but the increased need is well covered by a regular diet.

   a. true   b. false

8. The normal acid in the stomach is so strong that no food is acidic enough to make it stronger.

   a. true   b. false

9. Of the following foods, which contains amino acids that are best digested and absorbed?

   a. whole-wheat bread   c. legumes
   b. chicken   d. broccoli

10. The calorie deficiency disease is known specifically as:

    a. protein-energy malnutrition.   c. protein-calorie malnutrition.
    b. kwashiorkor.   d. marasmus.

11. Characteristics of children with marasmus include all of the following **except**:

    a. their digestive enzymes are in short supply.
    b. they retain some of their stores of body fat.
    c. all of their muscles are wasted.
    d. their metabolism is slowed.

12. Under normal circumstances, healthy adults are in:

    a. nitrogen equilibrium.   c. negative nitrogen balance.
    b. positive nitrogen balance.   d. minus zero balance.

13. Which of the following is in positive nitrogen balance?

    a. a growing child
    b. a person who rests in bed for a long time
    c. a pregnant woman
    d. a and b
    e. a and c

14. High-protein diets are associated with all of the following **except**:

    a. high-fat foods.   c. improved kidney function.
    b. increased risk of obesity.   d. increased risk of heart disease.

15. Health recommendations typically advise a protein intake between 10 and 35% of energy intake.

    a. true   b. false

16. Which of the following is **not** characteristic of legumes?

    a. They have nodules on their roots containing bacteria that can fix nitrogen.
    b. They capture nitrogen from the air and soil and use it to make amino acids.
    c. They are used by farmers in rotation with other crops to fertilize fields.
    d. Their protein quality is inferior to that of other plant foods.

17. Legumes are excellent sources of all of the following nutrients **except**:

   a. iron.
   b. calcium.
   c. vitamin C.
   d. protein.

18. Tofu is:

   a. naturally rich in calcium.
   b. a curd made from soybeans.
   c. a poor source of protein.
   d. high in fiber.

19. What food groups contribute an abundance of high-quality protein?

   a. meat, poultry, fish, legumes, eggs and nuts
   b. grains
   c. milk, yogurt and cheese
   d. a and b
   e. a and c

20. Under special circumstances, a nonessential amino acid can become essential.

   a. true
   b. false

21. The first step in the destruction of a protein is called:

   a. protein turnover.
   b. net protein utilization.
   c. denaturation.
   d. protein efficiency ratio.

22. Research supports the practice of long-term consumption of amino acid supplements by healthy people.

   a. true
   b. false

23. The millions of cells lining the intestinal tract live for _____ days.

   a. 1
   b. 2
   c. 3
   d. 4

24. The amino acid tryptophan serves as starting material for:

   a. the neurotransmitter serotonin.
   b. the hormone thyroxine.
   c. the vitamin niacin.
   d. a and b
   e. a and c

25. Amino acids from what source are the most easily digested and absorbed?

   a. legumes
   b. grains
   c. animals
   d. vegetables

Application Level Items:

26. Your best friend is a body builder who puts raw eggs in a milkshake to increase his protein intake. What argument would you use against this practice?

    a. Raw egg proteins bind the B vitamin biotin.
    b. Raw egg proteins speed up protein digestion.
    c. Raw egg proteins bind the mineral iron.
    d. a and b
    e. a and c

27. A person who centers his diet around protein is likely to receive an inadequate intake of:

    a. vitamin B_{12}.
    b. fat.
    c. folate.
    d. iron.

28. Valuable, expensive protein-rich foods can contribute to obesity when:

    a. there is a surplus of amino acids.
    b. they are used as is and become part of a growing protein.
    c. there is a surplus of energy-yielding nutrients.
    d. a and c
    e. b and c

29. Which of the following has the highest DRI for protein per unit of body weight?

    a. a 6-month-old infant
    b. a healthy 25-year-old female
    c. a 35-year-old male
    d. a 30-year-old female who exercises frequently

30. What advice would you give to a friend who wants to build bigger muscles?

    a. Consume a snack rich in protein and carbohydrate within 2 hours after exercise.
    b. Engage in rigorous physical training.
    c. Take an amino acid supplement.
    d. a and b
    e. b and c

31. Approximately the same amount of protein is provided by each of the following foods, which would be considered as cooked and/or ready to eat: ½ cup legumes; 1 ½ cups broccoli; 1 ounce cheese; 2 tablespoons peanut butter. Which of these would be the most nutrient-dense food choice?

    a. 1 ½ cups broccoli
    b. 2 tablespoons peanut butter
    c. 1 ounce cheese
    d. ½ cup legumes

32. What is the DRI for protein for a 50-year-old male weighing 198 pounds?

    a. 56
    b. 67
    c. 72
    d. 80

Answers

Summing Up

1. nitrogen
2. amino acids
3. twenty
4. shape
5. essential
6. peptide
7. acid
8. shapes
9. enzymes
10. acids
11. stomach acid
12. protein
13. mucus
14. small intestine
15. amino acids
16. cells
17. acid-base
18. carbohydrate
19. energy
20. amine
21. urea
22. animal
23. essential
24. protein
25. meat
26. soybean
27. plants
28. mutual supplementation
29. body weight
30. children
31. 0.8
32. PEM
33. marasmus
34. heart
35. milk
36. grain
37. minerals
38. folate

Chapter Glossary

Matching Exercise:

1. e
2. l
3. c
4. j
5. r
6. h
7. q
8. a
9. n
10. f
11. p
12. i
13. b
14. m
15. k
16. g
17. o
18. d

Crossword Puzzle:

1. tofu
2. alkalosis
3. legumes
4. textured vegetable protein
5. complementary proteins
6. marasmus
7. mutual supplementation
8. buffers
9. acidosis (across); amino acid (down)
10. urea
11. collagen
12. dysentery
13. limiting amino acid
14. serotonin
15. kwashiorkor

Exercises

Chapter Study Questions:

1. Many poisons are salts of heavy metals which denature proteins whenever they touch them. Milk is given so that the poison will act on the protein of the milk rather than on the protein tissues of the mouth, esophagus, and stomach.

2. Whenever there is not enough energy from carbohydrate or fat; when the diet's protein is low quality, with too few essential amino acids; when there is too much protein so that not all is needed; when there is too much of any single amino acid, such as from a supplement.

3. Marasmus reflects a chronic inadequate food intake and inadequate energy, vitamins, and minerals as well as too little protein; kwashiorkor results from severe acute malnutrition, with too little protein to support body functions.

4. Legumes are plants of the pea, bean and lentil family that have roots with nodules that contain special bacteria. These bacteria can trap nitrogen from the air in the soil and make it into compounds

that become part of the seed. The seeds are rich in high-quality protein compared with those of most other plants.

5. Cooking an egg denatures the protein and makes it firm; in addition, cooking denatures two raw-egg proteins that bind the B vitamin biotin and the mineral iron and slows protein digestion; therefore, cooking eggs liberates biotin and iron and aids digestion.

6. Because overconsumption of protein offers no benefits and may be related to health problems related to the heart, kidneys, and bones.

Short Answer Questions:

1. (a) oxygen; (b) hydrogen; (c) carbon; (d) nitrogen

2. (a) valine; (b) leucine; (c) isoleucine; (d) threonine; (e) lysine; (f) methionine; (g) phenylalanine; (h) tryptophan; (i) histidine

3. (a) used to build proteins; (b) converted to other small nitrogen-containing compounds such as niacin; (c) converted to some other amino acids; (d) converted to glucose or fat; (e) burned as fuel

4. (a) heat; (b) alcohol; (c) acids; (d) bases; (e) salts of heavy metals; (f) radiation

5. (a) meat; (b) fish; (c) poultry; (d) cheese; (e) eggs; (f) milk; (g) many soybean products

6. (a) growing child; (b) pregnant woman

7. (a) find body weight in pounds; (b) convert pounds to kilograms; (c) multiply by 0.8 g/kg to find total grams of protein per day

8. (a) meat; (b) milk; (c) vegetable; (d) grain

9. vitamin C; folate

10. B_{12}; iron

Problem-Solving:

1. 40 grams protein

2.

	Menu Item	Grams of Protein
Breakfast	1 cup skim milk	10
	1 boiled egg	6
	1 bagel	8
Lunch	2 oz. tuna fish	14
	2 slices whole-wheat bread	6
	2 tsp. mayonnaise	0
	½ cup strawberries	1
Dinner	2 oz. chicken breast	15
	½ cup broccoli	2
	1 slice whole-grain bread	3
	1 cup skim milk	10
	Total:	75

This menu provides almost double the amount of protein required by the 37-year-old female in problem #1.

Controversy Questions:

1. Some people do not believe in killing animals for their meat and believe that livestock is treated inhumanely. Others fear contracting diseases, such as mad cow disease, or practice vegetarianism for health reasons. Others may eat less meat over concern for the environment.

2. Obesity, high blood pressure, heart disease, and some forms of cancer.

3. Vegetarians typically use no tobacco, use alcohol in moderation and may be more physically active.

4. Alcohol, total food energy, well-cooked red meats and processed meats, and possibly high intakes of refined grain products.

5. Calcium, iron, zinc, vitamin D and vitamin B_{12} all require special attention for strict vegetarians.

Study Aid

1. (a) 4; (b) 3; (c) 6; (d) 1; (e) 5; (f) 7; (g) 2

2. (a) negative; (b) positive; (c) equilibrium; (d) positive; (e) negative

Sample Test Items

1. d (p. 186)
2. a (p. 186)
3. b (p. 189)
4. a (p. 199)
5. a (pp. 195-198)
6. b (p. 199)
7. a (p. 204)
8. a (p. 193)
9. b (p. 202)
10. d (p. 206)
11. b (pp. 206-207)
12. a (p. 205)
13. e (p. 205)
14. c (pp. 208-209)
15. a (p. 209)
16. d (pp. 210, 212)
17. c (p. 210)
18. b (p. 212)
19. e (p. 210)
20. a (p. 187)
21. c (p. 192)
22. b (p. 201)
23. c (p. 195)
24. e (p. 196)
25. c (p. 202)
26. e (p. 193)
27. c (p. 210)
28. d (pp. 199-200)
29. a (p. 204)
30. d (p. 192)
31. a (p. 211)
32. c (p. 204)

Chapter 7 - The Vitamins

Chapter Objectives

After completing this chapter, you should be able to:

1. Define the term vitamin and explain how vitamins are classified.

2. Describe the characteristics of fat-soluble vitamins and water-soluble vitamins and explain how they differ.

3. List and explain the chief functions, recommended intakes, food sources, and major deficiency and toxicity symptoms for each fat-soluble vitamin, including vitamin A, vitamin D, vitamin E and vitamin K.

4. Identify and explain the chief functions, recommended intakes, food sources, and mjaor deficiency and toxicity symptoms for the water-soluble vitamins, including vitamin C and the B vitamins: thiamin, riboflavin, niacin, folate, vitamin B_{12}, vitamin B_6, biotin and pantothenic acid.

5. Name the non-B vitamins and discuss what is known about them.

6. Describe the best method of discerning the vitamin content of foods and planning a diet that is rich in vitamins.

7. Discuss evidence regarding the pros and cons of vitamin supplements (Controversy 7).

8. Identify groups of people who may benefit from a multinutrient supplement and discuss guidelines for choosing an appropriate supplement (Controversy 7).

Key Concepts

✓ Vitamins are essential, noncaloric nutrients that are needed in tiny amounts in the diet and help to drive cell processes in the body. Vitamin precursors in foods are transformed into active vitamins in the body. The fat-soluble vitamins are vitamins A, D, E, and K; the water-soluble vitamins are vitamin C and the B vitamins.

✓ Vitamin A is essential to vision, integrity of epithelial tissue, bone growth, reproduction, and more. Vitamin A deficiency causes blindness, sickness, and death and is a major problem worldwide. Overdoses are possible and cause many serious symptoms. Foods are preferable to supplements for supplying vitamin A.

✓ The vitamin A precursor in plants, beta-carotene, is an effective antioxidant in the body. Brightly colored plant foods are richest in beta-carotene, and diets containing these foods are associated with good health.

✓ Vitamin D raises mineral levels in the blood, notably calcium and phosphorus, permitting bone formation and maintenance. A deficiency can cause rickets in childhood or osteomalacia in later life. Vitamin D is the most toxic of all the vitamins, and excesses are dangerous or deadly. People exposed to the sun make vitamin D from a cholesterol-like compound in their skin; fortified milk is an important food source.

✓ Vitamin E acts as an antioxidant in cell membranes and is especially important for the integrity of cells that are constantly exposed to high oxygen concentrations, namely, the lungs and red and white blood cells. Vitamin E deficiency is rare in human beings, but it does occur in newborn premature infants. The vitamin is widely distributed in plant foods; it is destroyed by high heat; toxicity is rare.

✓ Vitamin K is necessary for blood to clot; deficiency causes uncontrolled bleeding. The bacterial inhabitants of the digestive tract produce vitamin K. Toxicity causes jaundice.

✓ Vitamin C, an antioxidant, helps to maintain collagen, the protein of connective tissue; protects against infection; and helps in iron absorption. The theory that vitamin C prevents or cures colds or cancer is not well supported by research. Taking high vitamin C doses may be unwise. Ample vitamin C can be easily obtained from foods.

✓ As part of coenzymes, the B vitamins help enzymes do their jobs. The B vitamins facilitate the work of every cell. Some help generate energy; others help make protein and new cells. B vitamins work everywhere in the body tissues to metabolize carbohydrate, fat, and protein.

✓ Historically, famous B vitamin-deficiency diseases are beriberi (thiamin), pellagra (niacin), and pernicious anemia (vitamin B_{12}). Pellagra can be prevented by adequate protein because the amino acid tryptophan can be converted to niacin in the body. A high intake of folate can mask the blood symptom of vitamin B_{12} deficiency but will not prevent the associated nerve damage. Vitamin B_6 is important in amino acid metabolism and can be toxic in excess. Biotin and pantothenic acid are important to the body and are abundant in food.

✓ Choline is needed in the diet, but it is not a vitamin and deficiencies are unheard of outside the laboratory. Many other substances that people claim are B vitamins are not. Among these substances are carnitine, inositol, and lipoic acid.

Summing Up

A vitamin is defined as an essential, noncaloric, organic nutrient needed in (1)_____ amounts in the diet. The role of many vitamins is to help make possible the processes by which other nutrients are digested, (2)_____, and metabolized, or built into body (3)_____. Some of the vitamins occur in foods in a form known as (4)_____, or provitamins. The vitamins fall naturally into two classes including the (5)_____-soluble vitamins and the (6)_____-soluble vitamins.

Vitamin A plays parts in such diverse functions as gene expression, (7)_____, maintenance of skin and body linings, bone growth and reproduction. Active vitamin A is present in foods of (8)_____ origin. In plants, vitamin A exists only in its precursor forms; the most abundant of these is (9)_____. Rich food sources of beta carotene include carrots, sweet potatoes, pumpkins, mango, (10)_____, and apricots, as well as dark green vegetables.

Vitamin D is one member of a team of nutrients and hormones that maintains blood calcium and phosphorus levels and thereby (11)_____ integrity. The vitamin D deficiency disease in (12)_____ is called rickets, while the comparable disease in adults is referred to as (13)_____. Vitamin D is the most potentially (14)_____ of all vitamins and infants and older people taking vitamin D supplements are most often reported to develop toxicities. People can make vitamin D from a (15)_____ compound whenever the sun shines on their skin. The recommended intake for vitamin D is (16)_____ micrograms per day for adults 19 to 50 years.

Vitamin E is an (17)_____ and thus serves as one of the body's main defenders against oxidative damage. In the body, it exerts an especially important antioxidant effect in the (18)_____. Conditions that cause malabsorption of (19)_____ can cause vitamin E deficiencies, although vitamin E deficiency is rare in most situations. The DRI intake recommendation for vitamin E is (20)_____ milligrams a day for adults. Much of the vitamin E people consume comes from (21)_____ oils and products made from them.

Vitamin K is the fat-soluble vitamin necessary for the synthesis of proteins involved in (22)_____. It can be obtained from the nonfood source of (23)_____ bacteria. However, vitamin K deficiencies may occur in (24)_____ because they are born with sterile intestinal tracts.

The B vitamins and vitamin C are known together as the (25)_____ vitamins. The B vitamins function as part of (26)_____. Niacin, thiamin, riboflavin, pantothenic acid and biotin participate in the release of (27)_____ from carbohydrate, fat, and protein. For (28)_____, which is tied closely to amino acid metabolism, the amount needed is proportional to (29)_____ intake. The thiamin deficiency disease is called (30)_____, while the (31)_____ deficiency disease is known as pellagra. In the body, the amino acid (32)_____ is converted to niacin. To make new cells, tissues must have the vitamin (33)_____, which is important for reducing a pregnant woman's risk of having a child with birth defects known as (34)_____. Strict vegetarians are most apt to be deficient in (35)_____, which is present only in foods of animal origin. The vitamin C deficiency disease is scurvy, which can be prevented by an intake of (36)_____ milligrams of vitamin C per day. However, the adult DRI intake recommendation for vitamin C is (37)_____ milligrams for men and 75 milligrams for women.

Chapter Glossary

Matching Exercise 1:

_____ 1. biotin

_____ 2. vitamin B$_6$

_____ 3. rickets

_____ 4. erythrocyte hemolysis

_____ 5. choline

_____ 6. pellagra

_____ 7. scurvy

_____ 8. night blindness

_____ 9. prooxidant

_____ 10. niacin

_____ 11. riboflavin

_____ 12. vitamin B$_{12}$

_____ 13. inositol

_____ 14. ascorbic acid

_____ 15. beriberi

_____ 16. pernicious anemia

_____ 17. folate

_____ 18. thiamin

_____ 19. carotenoids

_____ 20. osteomalacia

a. the vitamin C-deficiency disease

b. a nonessential nutrient found in cell membranes

c. a B vitamin needed in protein metabolism; its three active forms are pyridoxine, pyridoxal, and pyridoxamine

d. a B vitamin needed in energy metabolism; it can be eaten preformed or can be made in the body from tryptophan

e. a B vitamin that acts as part of a coenzyme important in the manufacture of new cells

f. rupture of the red blood cells, caused by vitamin E deficiency

g. the thiamin-deficiency disease

h. a B vitamin; a coenzyme necessary for fat synthesis and other metabolic reactions

i. one of the active forms of vitamin C

j. the niacin-deficiency disease

k. a B vitamin active in the body's energy-release mechanisms

l. the vitamin D-deficiency disease in children

m. slow recovery of vision after exposure to flashes of bright light at night; an early symptom of vitamin A deficiency

n. a B vitamin that helps convert folate to its active form and also helps maintain the sheaths around nerve cells

o. a compound that triggers reactions involving oxygen

p. a nonessential nutrient used to make the phospholipid lecithin and other molecules

q. a vitamin B$_{12}$-deficiency disease, caused by lack of intrinsic factor and characterized by large, immature red blood cells and damage to the nervous system

r. a B vitamin involved in the body's use of fuels that occupies a special site on nerve cell membranes

s. the vitamin D deficiency disease in adults

t. a group of pigments in foods ranging from light yellow to reddish orange, many with vitamin A activity in the body

Matching Exercise 2:

_____ 1. serotonin

_____ 2. dietary antioxidants

_____ 3. carnitine

_____ 4. neural tube defects

_____ 5. epithelial tissue

_____ 6. keratinization

_____ 7. dietary folate equivalent (DFE)

_____ 8. macular degeneration

_____ 9. free radicals

_____ 10. niacin equivalents

_____ 11. pantothenic acid

_____ 12. jaundice

_____ 13. carpal tunnel syndrome

a. accumulation of keratin in a tissue; a sign of vitamin A deficiency

b. the amount of niacin present in food, including the niacin that can theoretically be made from its precursor tryptophan that is present in the food

c. yellowing of the skin due to spillover of the bile pigment bilirubin from the liver into the general circulation

d. a neurotransmitter important in sleep regulation, appetite control, and mood regulation, among other roles

e. abnormalities of the brain and spinal cord apparent at birth and believed to be related to a woman's folate intake before and during pregnancy

f. a B vitamin

g. compounds typically found in plant foods that significantly decrease the adverse effects of oxidation on living tissues

h. the layers of the body that serve as selective barriers to environmental factors

i. a unit of measure expressing the amount of folate available to the body from naturally occurring sources

j. a nonessential nutrient that functions in cellular activities

k. a common, progressive loss of function of the part of the retina that is most crucial to focused vision; often leads to blindness

l. atoms or molecules with one or more unpaired electrons that make the atom or molecule unstable and highly reactive

m. a pinched nerve at the wrist, causing pain or numbness in the hand; it is often caused by repetitive motion of the wrist

Crossword Puzzle:

Across:

1. the normal protein of hair and nails
4. hardening of the cornea of the eye in advanced vitamin A deficiency that can lead to blindness
9. a small molecule that works with an enzyme to promote the enzyme's activity
10. the light-sensitive pigment of the cells in the retina
12. a factor found inside a system
14. one of the active forms of vitamin A made from beta-carotene in animal and human bodies
15. the chief protein of most connective tissues, including scars, ligaments, and tendons, and the underlying matrix on which bones and teeth are built
16. a weakening of bone mineral structures that occurs commonly with advancing age

Down:

2. the layer of light-sensitive nerve cells lining the back of the inside of the eye
3. an orange pigment with antioxidant activity; a vitamin A precursor made by plants and stored in human fat tissue
5. compounds that can be converted into active vitamins
6. organic compounds that are vital to life and indispensable to body functions, but are needed only in minute amounts
7. drying of the cornea; a symptom of vitamin A deficiency
8. the hard, transparent membrane covering the outside of the eye
11. a kind of alcohol
13. a measure of fat-soluble vitamin activity sometimes used on supplement labels

Word Bank:

beta-carotene	intrinsic factor	precursors	tocopherol
coenzyme	IU	retina	vitamins
collagen	keratin	retinol	xerophthalmia
cornea	osteoporosis	rhodopsin	xerosis

Exercises

Answer these chapter study questions:

1. List the general characteristics of fat-soluble vitamins.

2. Do individuals have to eat vitamin D to have enough in their bodies? Why or why not?

3. Why do diseases or injuries which compromise the liver, gallbladder or pancreas make vitamin E deficiency likely?

4. Why would someone who had part of the stomach removed develop deficiency symptoms for vitamin B_{12}?

5. Is it possible for a person consuming adequate protein and adequate calories to be deficient in niacin? Why or why not?

6. What has been responsible for the 25 percent drop in the national incidence of neural tube defects?

Complete these short answer questions:

1. The fat-soluble vitamins include:

    a.   c.

    b.   d.

2. Vitamin A has parts to play in such diverse functions as:

    a.

    b.

    c.

    d.

    e.

    f.

    g.

3. The richest food sources of beta-carotene include:

    a.   e.

    b.   f.

    c.   g.

    d.   h.

4. The earliest symptoms of vitamin A overdoses are:

    a.   d.

    b.   e.

    c.   f.

5. To raise the level of blood calcium, the body can draw from these places:

    a.

    b.

    c.

6. Vitamin K's richest plant sources are:

    a.

    b.

7. These individuals may incur, over a period of years, a vitamin D deficiency:

    a.

    b.

    c.

8. In adults, vitamin E deficiency is usually associated with diseases of these organs:

    a.   c.

    b.

9. The best natural sources of folate are _____ vegetables and fruits.

10. In the U.S. today, scurvy may be seen in these groups:

    a.

    b.

    c.

Solve these problems:

1. What is the impact on water-soluble vitamins when foods are cut, washed, and then cooked?

2. If a person consumed a diet composed of 1200 micrograms of beta-carotene from food, how many micrograms of retinol would this supply in the body?

Answer these controversy questions:

1. Describe the law regarding supplements that went into effect in 2007.

2. A supplement manufacturer claims that today's foods lack sufficient nutrients to support health. How would you respond to that claim?

3. What is the primary characteristic of the diet of populations with low cancer rates?

4. What is a potential harmful effect of taking a vitamin E supplement?

5. You see a USP symbol on the label of a supplement. What does this symbol tell the consumer?

Study Aids

1. Complete the following chart which identifies names, chief functions, deficiency disease names, and major food sources of some of the fat-soluble and water-soluble vitamins.

Name	Chief Functions	Deficiency Disease Name(s)	Food Sources
A	• _____ • bone _____ • reproduction • _____ cells	_____	*For retinol:* • fortified _____, cheese, cream, butter • _____ margarine • eggs • _____ *For beta-carotene:* • dark leafy _____ • deep orange fruits and _____
D	• mineralization of _____	rickets; _____	• _____ milk or margarine • sardines • liver • shrimp • _____
E	• protects PUFA • normal _____ development • support of _____ function	no name	• polyunsaturated plant _____ • _____ germ • green leafy vegetables

Name	Chief Functions	Deficiency Disease Name(s)	Food Sources
K	• synthesis of blood _____ proteins	no name	• _____ • _____-t _____ vegetables type • green _____ vegetables • _____ oils • occurs in all nutritious foods
Thiamin	• part of a _____ used in _____ energy metabolism	_____	
Riboflavin	• part of a coenzyme used in _____ metabolism	_____	• milk, yogurt and cottage cheese • meat and _____ • leafy green vegetables • whole-grain or _____ breads and
Niacin	• part of a _____ used in _____ energy metabolism	_____	• milk • meat • _____ • fish • enriched and _____ and cereals • all _____-containing foods

Name	Chief Functions	Deficiency Disease Name(s)	Food Sources
Vitamin C	• _____ s _____ ynthesis • _____ • hormone synthesis • supports _____ cell functions • restores _____ to active form • helps in absorption of _____	_____	• _____ fruits • _____ -t and dark green vegetables • _____ cantaloupe • _____ peppers • lettuce • _____ potatoes • _____ or mangoes

2. Use Table C7-1 on page 261 of your textbook to study valid reasons for taking supplements.

Sample Test Items

Comprehension Level Items:

1. The vitamins fall naturally into _____ classes.

   a. two
   b. three
   c. four
   d. five

2. All of the following are characteristics of fat-soluble vitamins **except**:

   a. they require bile for absorption.
   b. they can reach toxic levels.
   c. they are easily excreted.
   d. they are stored in fatty tissues.

3. The first fat-soluble vitamin to be recognized was vitamin:

   a. D.
   b. K.
   c. E.
   d. A.

4. A vitamin A precursor:

   a. is called beta-carotene.
   b. is found in plants.
   c. can cause vitamin A toxicity.
   d. a and b
   e. a and c

5. All of the following are good sources of beta-carotene **except**:

   a. carrots.
   b. pumpkins.
   c. red cabbage.
   d. apricots.

6. The vitamin D deficiency disease in children is known as:

   a. osteodystrophy.
   b. rickets.
   c. pellagra.
   d. osteomalacia.

7. People who are exposed to the sun make vitamin D from:

   a. cholesterol.
   b. carotene.
   c. polyunsaturated fats.
   d. saturated fats.

8. Vitamin E exerts an especially important antioxidant effect in the:

   a. kidneys.
   b. heart.
   c. lungs.
   d. stomach.

9. The DRI for vitamin E is _____ milligrams a day for adults.

   a. 5
   b. 8
   c. 10
   d. 15

10. Vitamin K deficiencies may occur in:

    a. newborn infants.
    b. people who have undergone surgery.
    c. people who have taken antibiotics.
    d. a and c
    e. b and c

11. The niacin deficiency disease is known as:

    a. osteomalacia.
    b. scurvy.
    c. pellagra.
    d. beriberi.

12. A high intake of _____ can mask the anemia caused by vitamin B_{12} deficiency.

    a. thiamin
    b. folate
    c. riboflavin
    d. niacin

13. What is the amount of vitamin C required to prevent the symptoms of scurvy from appearing?

    a. 10 milligrams
    b. 20 milligrams
    c. 30 milligrams
    d. 40 milligrams

14. For which vitamin is the amount needed proportional to protein intake?

    a. thiamin
    b. riboflavin
    c. niacin
    d. vitamin B_6

15. People addicted to alcohol may develop a deficiency of:

    a. niacin.
    b. thiamin.
    c. vitamin B_6.
    d. vitamin B_{12}.

16. Vitamin B_{12} is naturally occurring only in:

    a. grains.
    b. fruits.
    c. animals.
    d. vegetables.

17. In the U.S. today, scurvy has been found in all of the following **except**

    a. breast-fed infants.
    b. infants fed only cow's milk.
    c. elderly.
    d. people addicted to drugs.

18. Between 3 and 10 million of the world's children suffer from severe vitamin _____ deficiency.

    a. C
    b. A
    c. B_{12}
    d. D

19. Which of the following supplements would be the most appropriate to choose?

    a. one that comes in a chewable form
    b. one that provides all the vitamins and minerals in amounts smaller than or close to the intake recommendations
    c. one that provides more than the DRI recommended intake for vitamin A, vitamin D and minerals
    d. one that provides more than the Tolerable Upper Intake Level for calcium

20. Vitamin C supplements in any dosage may be dangerous for people with an overload of _____ in the blood.

    a. zinc
    b. calcium
    c. magnesium
    d. iron

21. The adult DRI intake recommendation for vitamin C is _____ milligrams for women.

    a. 75   c. 105
    b. 90   d. 125

22. Which of the following has gained a reputation as an "anti-infective" vitamin?

    a. D   c. E
    b. A   d. K

23. Which of the following food sources of vitamin A should **not** be eaten on a daily basis because of concerns about toxicity?

    a. fish oil   c. liver
    b. eggs   d. butter

24. The most potentially toxic of all vitamins is vitamin:

    a. A.   c. D.
    b. C.   d. K.

25. Diseases of the liver, gallbladder, and pancreas are likely to result in a deficiency of vitamin

    a. A.   c. K.
    b. C.   d. E.

Application Level Items:

26. A friend of yours complains that she is unable to see for a short period of time after she encounters a flash of bright light at night. Which of the following statements would you make to your friend?

    a. "Your eyes are probably well nourished and normal."
    b. "You are suffering from vitamin A deficiency."
    c. "You have night blindness."
    d. "You may need to check your diet for vitamin A deficiency."

27. A couple you know gives their infant a jar of baby carrots to eat every day. This child runs the risk of:

    a. developing an allergic reaction to carrots.
    b. developing a vitamin A toxicity.
    c. having his skin turn bright yellow.
    d. a and b
    e. b and c

28. Which of the following would have the lowest need for vitamin A?

    a. a man   c. a woman
    b. a child   d. a lactating woman

29. You have a friend who has been experiencing appetite loss, nausea, vomiting, and increased urination and thirst. A medical examination reveals that calcium has been deposited in the heart and kidneys. Based on these symptoms you suspect that your friend has been taking supplements of:

    a. vitamin A.   c. vitamin D.
    b. vitamin K.   d. vitamin E.

30. Which of the following would **not** be at risk for developing a vitamin D deficiency?

    a. an elderly woman in a nursing home
    b. a man who works the night shift for several years
    c. a dark-skinned child who lives in a northern city
    d. an adult who eats cereal with milk for breakfast every day

31. An athlete who consumes a high-protein diet would have an increased need for:

    a. vitamin B_6.
    b. vitamin C.
    c. vitamin B_{12}.
    d. thiamin.

32. You are reading the labels of four vitamin supplements and see one of the words or symbols listed below on each label. Which supplement would you choose?

    a. organic
    b. USP symbol
    c. stress formula
    d. inositol

Answers

Summing Up

1. tiny
2. absorbed
3. structures
4. precursors
5. fat
6. water
7. vision
8. animal
9. beta-carotene
10. cantaloupe
11. bone
12. children
13. osteomalacia
14. toxic
15. cholesterol
16. 5
17. antioxidant
18. lungs
19. fat
20. 15
21. vegetable
22. blood clotting
23. intestinal
24. infants
25. water-soluble
26. coenzymes
27. energy
28. vitamin B_6
29. protein
30. beriberi
31. niacin
32. tryptophan
33. folate
34. neural tube defects
35. vitamin B_{12}
36. 10
37. 90

Chapter Glossary

Matching Exercise 1:
1. h
2. c
3. l
4. f
5. p
6. j
7. a
8. m
9. o
10. d
11. k
12. n
13. b
14. i
15. g
16. q
17. e
18. r
19. t
20. s

Matching Exercise 2:
1. d
2. g
3. j
4. e
5. h
6. a
7. i
8. k
9. l
10. b
11. f
12. c
13. m

Crossword Puzzle:

1. keratin
2. retina
3. beta-carotene
4. xerophthalmia
5. precursors
6. vitamins
7. xerosis
8. cornea
9. coenzyme
10. rhodopsin
11. tocopherol
12. intrinsic factor
13. IU
14. retinol
15. collagen
16. osteoporosis

Exercises

Chapter Study Questions:

1. They require bile for absorption. They generally occur together in the fats and oils of foods; once they have been absorbed from the intestinal tract, they are stored in the liver and fatty tissues until the body needs them; excesses can reach toxic levels; they can be lost from the digestive tract with undigested fat; they play diverse roles in the body.

2. Humans do not have to eat vitamin D because they can make it from a cholesterol compound whenever ultraviolet light from the sun shines on their skin. The compound is transformed into a vitamin D precursor and is absorbed directly into the blood.

3. Because the liver makes bile which is necessary for the digestion of fat; the gallbladder stores bile and delivers it into the intestine; and the pancreas makes fat-digesting enzymes.

4. The absorption of vitamin B_{12} requires an intrinsic factor, which is synthesized in the stomach. When part of the stomach has been removed, it cannot produce enough of the intrinsic factor. Without enough of the intrinsic factor, the person cannot absorb the vitamin even though he may be getting enough in the diet.

5. A person eating adequate protein will not be deficient in niacin because the amino acid tryptophan is converted to niacin in the body.

6. In the late 1990s, the FDA ordered fortification of all enriched grain products with folic acid, which has resulted in dramatic increases in folate intakes. Folate deficiency is associated with neural tube defects.

Short Answer Questions:

1. (a) A; (b) D; (c) E; (d) K

2. (a) vision; (b) maintenance of body linings and skin; (c) bone and body growth; (d) reproduction; (e) normal cell development; (f) immune defenses; (g) gene expression

3. (a) carrots; (b) sweet potatoes; (c) pumpkins; (d) cantaloupe; (e) apricots; (f) spinach; (g) broccoli; (h) mango

4. (a) blurred vision; (b) dizziness; (c) loss of appetite; (d) headache; (e) itching of skin; (f) irritability

5. (a) skeleton; (b) food containing calcium in the digestive tract; (c) kidneys, which recycle calcium that would otherwise be lost in urine

6. (a) dark-green leafy vegetables; (b) members of the cabbage family

7. (a) people who are housebound or institutionalized; (b) those who work at night; (c) dark-skinned people who live in smoggy northern cities

8. (a) liver; (b) gallbladder; (c) pancreas

9. uncooked

10. (a) infants fed only cow's milk; (b) elderly; (c) people addicted to alcohol or other drugs

Problem-Solving:

1. The water-soluble vitamins, including vitamin C and the B vitamins, may leach out of the foods because they dissolve in water.

2. 100 micrograms of retinol

Controversy Questions:

1. The law requires that manufacturers report to FDA any deaths, hospitalizations, life-threatening events, birth defects, disabilities or medical interventions resulting from their supplements and reported to them by consumers.

2. Plants make vitamins to meet their needs. A plant that lacks a mineral or fails to make a needed vitamin will die before it can bear food for human consumption.

3. They consume abundant vegetables and fruits, especially those containing antioxidants, such as beta-carotene.

4. Vitamin E supplements resulted in an increased risk of death when results from high-quality studies were pooled together.

5. The USP symbol means that a manufacturer has voluntarily paid an independent laboratory to test the product and affirm that it contains the ingredients as listed and that it will dissolve or disintegrate in the digestive tract to make the ingredients available for absorption.

Study Aids

Name	Chief Functions	Deficiency Disease Name(s)	Food Sources
A	• vision • bone growth • reproduction • epithelial cells	hypovitaminosis A	*For retinol:* • fortified milk, cheese, cream and butter • fortified margarine • eggs • liver *For beta-carotene:* • dark leafy greens • deep orange fruits and vegetables
D	• mineralization of bones	rickets; osteomalacia	• fortified milk or margarine • sardines • liver • shrimp • salmon

Name	Chief Functions	Deficiency Disease Name(s)	Food Sources
E	• antioxidant • protects PUFA • normal nerve development • support of immune function	no name	• polyunsaturated plant oils • wheat germ • green, leafy vegetables
K	• synthesis of blood clotting proteins	no name	• soybeans • cabbage-type vegetables • green leafy vegetables • vegetable oils
Thiamin	• part of a coenzyme used in energy metabolism	beriberi	• occurs in all nutritious foods
Riboflavin	• part of a coenzyme used in energy metabolism	ariboflavinosis	• milk, yogurt and cottage cheese • meat and liver • leafy green vegetables • whole-grain or enriched breads and cereals
Niacin	• part of a coenzyme used in energy metabolism	pellagra	• milk • eggs • meat • poultry • fish • enriched and whole-grain breads and cereals • all protein-containing foods
Vitamin C	• collagen synthesis • antioxidant • hormone synthesis • supports immune cell functions • restores vitamin E to active form • helps in absorption of iron	scurvy	• citrus fruits • cabbage-type and dark green vegetables • cantaloupe • strawberries • peppers • lettuce • tomatoes • potatoes • papayas or mangoes

Sample Test Items

1. a (p. 220)
2. c (pp. 220-221)
3. d (p. 222)
4. d (p. 227)
5. c (p. 227)
6. b (p. 228)
7. a (p. 229)
8. c (p. 232)
9. d (p. 232)
10. d (p. 234)
11. c (p. 244)
12. b (p. 247)
13. a (pp. 237-238)
14. d (p. 250)
15. b (p. 242)
16. c (p. 248)
17. a (p. 237)
18. b (p. 224)
19. b (pp. 266-268)
20. d (p. 238)
21. a (p. 238)
22. b (p. 224)
23. c (p. 225)
24. c (p. 229)
25. d (p. 232)
26. a (p. 223)
27. c (p. 227)
28. b (pp. 225-226)
29. c (p. 229)
30. d (pp. 229-231)
31. a (p. 250)
32. b (p. 266)

120

Chapter 8 - Water and Minerals

Chapter Objectives

After completing this chapter, you should be able to:

1. Discuss the major roles of water in the body.

2. Describe how the body regulates water intake and excretion to maintain water balance.

3. State the amount of water needed by adults and factors that influence the need for water.

4. Describe systems in place for ensuring that public water is safe to drink.

5. Compare the arguments for and against drinking bottled water.

6. Explain the role of minerals in maintaining the body's fluid and electrolyte balance and acid-base balance.

7. List the major roles for each major mineral and discuss recommended intakes, good food sources, and deficiency and toxicity symptoms.

8. List the major roles for each trace mineral and discuss recommended intakes, good food sources, and deficiency and toxicity symptoms.

9. Discuss the lifestyle choices that reduce the risk of osteoporosis (Controversy 8).

Key Concepts

✓ Water provides the medium for transportation, acts as a solvent, participates in chemical reactions, provides lubrication and shock protection, and aids in temperature regulation in the human body.

✓ Water makes up about 60 percent of the body's weight. A change in the body's water content can bring about a temporary change in body weight.

✓ Water losses from the body necessitate intake equal to output to maintain balance. The brain regulates water intake; the brain and kidneys regulate water excretion. Dehydration and water intoxication can have serious consequences.

✓ Many factors influence a person's need for water. The water of beverages and foods meets nearly all of the need for water, and a little more is supplied by the water formed during cellular breakdown of energy nutrients.

✓ Hard water is high in calcium and magnesium. Soft water is high in sodium, and it dissolves cadmium and lead from pipes.

✓ Public drinking water is tested and treated for safety. All drinking water originates from surface water or ground water that is vulnerable to contamination from human activities.

✓ Mineral salts form electrolytes that help keep fluids in their proper compartments and buffer these fluids, permitting all life processes to take place.

✓ Calcium makes up bone and tooth structure and plays roles in nerve transmission, muscle contraction, and blood clotting. Calcium absorption rises when there is a dietary deficiency or an increased need such as during growth.

✓ Most of the phosphorus in the body is in the bones and teeth. Phosphorus helps maintain acid-base balance, is part of the genetic material in cells, assists in energy metabolism, and forms part of cell membranes. Under normal circumstances, deficiencies of phosphorus are unknown.

✓ Most of the body's magnesium is in the bones and can be drawn out for all the cells to use in building protein and using energy. Most people in the United States choose diets that lack sufficient magnesium.

✓ Sodium is the main positively charged ion outside the body's cells. Sodium attracts water. Thus, too much sodium (or salt) raises blood pressure and aggravates hypertension. Diets rarely lack sodium.

✓ Potassium, the major positive ion inside cells, is important in many metabolic functions. Fresh whole foods are the best sources of potassium. Diuretics can deplete the body's potassium and so can be dangerous; potassium excess can also be dangerous.

✓ Chloride is the body's major negative ion; it is responsible for stomach acidity and assists in maintaining proper body chemistry. No known diet lacks chloride.

✓ Sulfate is a necessary nutrient used to synthesize sulfur-containing body compounds.

✓ Iodine is part of the hormone thyroxine, which influences energy metabolism. The deficiency diseases are goiter and cretinism. Iodine occurs naturally in seafood and in foods grown on land that was once covered by oceans; it is an additive in milk and bakery products. Large amounts are poisonous. Potassium iodide, appropriately administered, blocks some radiation damage to the thyroid during radiation emergencies.

✓ Most iron in the body is contained in hemoglobin and myoglobin or occurs as part of enzymes in the energy-yielding pathways. Iron-deficiency anemia is a problem worldwide; too much iron is toxic. Iron is lost through menstruation and other bleeding; reduced absorption and the shedding of intestinal cells protect against overload. For maximum iron absorption, use meat, other iron sources, and vitamin C together.

✓ Zinc assists enzymes in all cells. Deficiencies in children cause growth retardation with sexual immaturity. Zinc supplements can reach toxic doses, but zinc in foods is nontoxic. Foods from animals are the best sources.

✓ Selenium works with an enzyme system to protect body compounds from oxidation. A deficiency induces a disease of the heart. Deficiencies are rare in developed countries, but toxicities can occur from overuse of supplements.

✓ Fluoride stabilizes bones and makes teeth resistant to decay. Excess fluoride discolors teeth; large doses are toxic.

✓ Chromium works with the hormone insulin to control blood glucose concentrations. Chromium is present in a variety of unrefined foods.

✓ Copper is needed to form hemoglobin and collagen and assists in many other body processes. Copper deficiency is rare.

✓ Many different trace elements play important roles in the body. All of the trace minerals are toxic in excess.

Summing Up

Water makes up about (1)_____ percent of all the body's weight and is the most indispensable nutrient of all. Thirst and (2)_____ govern water intake, although (3)_____ lags behind a lack of water. Water excretion is governed by the brain and the (4)_____, while the (5)_____ regulates water intake. A change in the body's water content can bring a change in body (6)_____. Water losses from the body necessitate intake of an equal amount of water to maintain water (7)_____. In the body, water provides the medium for (8)_____, chemical reactions, shock protection, lubrication, and (9)_____ regulation.

Under normal dietary and environmental conditions, men need about (10)_____ cups of fluid from beverages and drinking water and women need about 9 cups. This amount of fluid provides about (11)_____ percent of the day's need for water. Hard water is high in calcium and (12)_____, while soft water is high in (13)_____.

Most of the body's water weight is contained inside the (14)_____, and some water also bathes the (15)_____ of the cells. Body cells can pump (16)_____ across their membranes, but not water, and these minerals attract the water to come along with them. The cells use minerals for this purpose in a special form, known as ions or (17)_____. The result of the system's working properly is (18)_____ and electrolyte balance. The minerals also help manage another balancing act, the (19)_____ balance. Among the major minerals, some, when dissolved in water, give rise to acids, and some to (20)_____.

Calcium is by far the most (21)_____ mineral in the body. Calcium makes up bone and (22)_____ structure and plays roles in (23)_____ transmission, muscle contraction and blood clotting. Adult bone loss is called (24)_____. The DRI recommended intakes for calcium are high for children and adolescents, because people develop their (25)_____ during their growing years.

Sodium is the chief ion used to maintain the volume of fluid (26)_____ cells and it contributes (27)_____ percent of the weight of the compound sodium chloride. Sodium helps maintain fluid and electrolyte balance and (28)_____ balance and is essential to nerve transmission and muscle contraction. The DRI intake recommendation for sodium has been set at (29)_____ milligrams for healthy, active young adults.

Both potassium deficiency and excess can be dangerous and (30)_____ can deplete body potassium. Other major minerals include (31)_____, which helps strands of protein to assume and hold their functional shapes, magnesium, phosphorus and (32)_____, which is part of the stomach's hydrochloric acid.

One of the trace minerals is iodine, which is part of the hormone (33)_____, which regulates the basal metabolic rate. Every living cell contains (34)_____, which is contained in hemoglobin in red blood cells and (35)_____ in muscle cells. Zinc works with proteins in every organ as a helper for nearly 100 (36)_____ and the best sources are animal foods.

Fluoride stabilizes bone and makes (37)_____ resistant to decay.

Chapter Glossary

Matching Exercise:

_____	1. hemoglobin	a. compounds in tea (especially black tea), and coffee that bind iron
_____	2. osteoporosis	b. enlargement of the thyroid gland due to iodine deficiency
		c. the oxygen-holding protein of the muscles
_____	3. fluid and electrolyte balance	d. the highest attainable bone density for an individual
		e. the oxygen-carrying protein of the blood
_____	4. goiter	f. severe mental and physical retardation of an infant caused by the mother's iodine deficiency during her pregnancy
_____	5. phytates	g. maintenance of the proper amounts and kinds of fluids and minerals in each compartment of the body
_____	6. peak bone mass	h. the condition of having depleted iron stores, which, at the extreme, causes iron-deficiency anemia
_____	7. iron overload	i. maintenance of the proper degree of acidity in each of the body's fluids
_____	8. buffers	
_____	9. tannins	j. a reduction of the bone mass of older persons in which the bones become porous and fragile
_____	10. heme	k. molecules that can help to keep the pH of a solution from changing by gathering or releasing H ions
_____	11. dehydration	l. the chief crystal of bone, formed from calcium and phosphorus
_____	12. iron deficiency	m. loss of water
		n. a craving for nonfood substances
_____	13. hydroxyapatite	o. the state of having more iron in the body than it needs or can handle, usually from a hereditary defect
_____	14. cretinism	
_____	15. MFP factor	p. a factor (identity unknown) present in meat, fish, and poultry that enhances the absorption of nonheme iron present in the same foods or in other foods eaten at the same time
_____	16. acid-base balance	
_____	17. myoglobin	q. the iron-containing portion of the hemoglobin and myoglobin molecules
_____	18. pica	r. compounds present in plant foods (particularly whole grains) that bind iron and prevent its absorption
_____	19. nori	s. a yogurt-based beverage
_____	20. kefir	t. a type of seaweed popular in Asian, particularly Japanese, cooking

Crossword Puzzle:

Across:	Down:
2. a dangerous dilution of the body's fluids, usually from excessive ingestion of plain water	1. essential mineral nutrients found in the human body in amounts larger than 5 grams
3. a crystal of bones and teeth, formed when fluoride displaces the hydroxy portion of hydroxyapatite	4. compounds that partly dissociate in water to form ions, such as the potassium ion (K^+) and the chloride ion (Cl^-)
5. the condition of inadequate or impaired red blood cells	6. naturally occurring, inorganic, homogenous substances; chemical elements
7. compounds, usually medications, causing increased urinary water excretion	11. literally, "lightened" by yeast cells, which digest some carbohydrate components of the dough and leave behind bubbles of gas that make the bread rise
8. the balance between water intake and water excretion, which keeps the body's water content constant	12. compounds composed of charged particles
9. essential mineral nutrients found in the human body in amounts less than 5 grams	
10. discoloration of the teeth due to ingestion of too much fluoride during tooth development	
12. water with a high sodium concentration	
13. electrically charged particles, such as sodium (positively charged) or chloride (negatively charged)	
14. high blood pressure	
15. water with high calcium and magnesium concentrations	

Exercises

Answer these chapter study questions:

1. Discuss the health implications of drinking hard water versus soft water.

2. Why is setting recommended intakes for calcium difficult?

3. Why are recommended intakes for calcium higher for children and adolescents?

4. Why should someone taking diuretics, which cause potassium loss, be advised to eat potassium-rich foods?

5. Define the term *pica*. What type of mineral deficiency is associated with pica?

6. What does *DASH* stand for, and what are the characteristics of the DASH diet?

Complete these short answer questions:

1. The major minerals include:

   a.   e.

   b.   f.

   c.   g.

   d.

2. The distinction between hard water and soft water is based on these three minerals:

    a.

    c.

    b.

3. The major roles of calcium in the body's fluids include:

    a.

    b.

    c.

    d.

    e.

    f.

4. All of the following can deplete the body of needed sodium:

    a.

    b.

    c.

    d.

5. The person who wishes to learn to control salt intake should:

    a.

    b.

6. Deficiency of magnesium may occur as a result of:

    a.

    d.

    b.

    e.

    c.

7. The need for iodine is easily met by consuming:

    a.

    b.

8. Substances that hinder iron absorption include:

    a.

    b.

    c.

Solve these problems:

Modify the following recipe to make it lower in calories, fat, and sodium and to substitute polyunsaturated fat for saturated fat.

Chicken Pot Pie	**Modifications:**
Pastry	
1 cup flour	
¼ tsp. salt	
2 tbsp. water	
$1/3$ cup lard	
Filling	
$1/3$ cup chopped potatoes	
$1/3$ cup sliced carrots	
$1/3$ cup canned green peas	
¼ cup chopped celery	
1 tbsp. chopped onion	
½ cup boiling water	
¼ cup butter	
¼ cup flour	
½ tsp. salt	
$1/8$ tsp. pepper	
$1/8$ tsp. poultry seasoning	
$1 1/3$ cup chicken broth	
$2/3$ cup cream	
1 ½ cup diced, cooked chicken	

Answer these controversy questions:

1. Differentiate between trabecular bone and cortical bone.

2. What are environmental factors currently under study for their roles in lowering bone density?

3. How is genetics thought to influence bone density?

4. What types of physical activity are recommended for building muscle and bone strength to reduce the risk of osteoporosis?

5. What are the guidelines for growing healthy bones during childhood?

Study Aids

1. Complete the following chart which identifies the names, chief functions, deficiency disease names, and major food sources of some of the major and trace minerals.

Major Minerals			
Name	Chief Functions	Deficiency Disease Name	Food Sources
Calcium	• makes up bone and _____ structure • normal muscle contraction • nerve functioning • blood _____ • immune defenses • blood pressure	_____	• milk and milk products • small _____ with bones • tofu • certain leafy greens • legumes

Major Minerals			
Name	Chief Functions	Deficiency Disease Name	Food Sources
Magnesium	• _____ mineralization • building of protein • normal muscle contraction • transmission of nerve • maintenance of _____		• nuts • _____ • whole grains • dark green _____ • seafoods • _____ • cocoa
Sodium	• maintains normal fluid balance • maintains normal _____ balance		• salt • _____ foods • _____ sauce

Trace Minerals			
Name	Chief Functions	Deficiency Disease Name	Food Sources
Iron	• part of the protein _____ • part of the protein myoglobin • necessary for utilization of _____ in the body	_____	• _____ meats • _____ • poultry • shellfish • eggs • legumes • _____ fruits
Iodine	• component of thyroid hormone _____	_____ _____	• _____ • _____ salt • bread
Zinc	• part of insulin and many _____ • normal _____ development		• _____-containing foods

2. Study the sodium and salt intake guidelines in Table 8-5 on page 287 of your textbook.

Sample Test Items

Comprehension Level Items:

1. The distinction between the major and trace minerals is that:

   a. the major minerals are more important than the trace minerals.
   b. the major minerals play more important roles than the trace minerals.
   c. the major minerals are present in larger quantities than the trace minerals.
   d. a and c
   e. b and c

2. All of the following play a role in water intake **except**

   a. the mouth.
   b. the volume of blood.
   c. the hypothalamus.
   d. the lungs.

3. One of the principal minerals in hard water is:

   a. calcium.
   b. fluoride.
   c. sodium.
   d. potassium.

4. Water needs vary greatly depending on:

   a. the foods a person eats.
   b. the environmental temperature and humidity.
   c. the person's gender.
   d. a and b
   e. b and c

5. The body's proteins and some of its mineral salts prevent changes in the acid-base balance of its fluids by serving as:

   a. ions.
   b. electrolytes.
   c. buffers.
   d. solutes.

6. Adult bone loss is referred to as:

   a. rickets.
   b. osteomalacia.
   c. osteodystrophy.
   d. osteoporosis.

7. Which of the following is(are) characteristic of osteoporosis?

   a. bones become porous
   b. bones become fragile
   c. bones bend
   d. a and b
   e. b and c

8. Which of the following groups absorbs the greatest amount of calcium?

   a. infants
   b. pregnant women
   c. adolescents
   d. older adults

9. Which of the following supplies most of the calcium to the U.S. diet?

   a. butter and cream
   b. small fish with bones

   c. tofu
   d. milk and milk products

10. The DRI Committee has set a Tolerable Upper Intake Level for sodium of _____ milligrams per day.

   a. 300
   b. 500

   c. 2300
   d. 3200

11. Processed and fast foods are the source of almost _____ percent of the salt in most people's diets.

   a. 15
   b. 35

   c. 52
   d. 75

12. The most productive step to take to control salt intake is to:

   a. stop using instant foods.
   b. discontinue the use of milk products.
   c. limit intake of processed and fast foods.
   d. control use of the salt shaker.

13. Which of the following can cause potassium depletion?

   a. use of diuretics that cause potassium loss
   b. severe diarrhea
   c. excessive use of potassium chloride
   d. a and b
   e. b and c

14. Which of the following is part of the hydrochloric acid that maintains the strong acidity of the stomach?

   a. chloride
   b. sulfur

   c. magnesium
   d. potassium

15. Foods rich in magnesium are:

   a. slightly processed.
   b. unprocessed.
   c. highly processed.

   d. a and c
   e. a and b

16. Which of the following is **not** a completely dependable source of iodine?

   a. seafood
   b. sea salt

   c. iodized salt
   d. plants grown in iodine-rich soil

17. All of the following are symptoms of iron deficiency **except**:

   a. tiredness.
   b. impaired concentration.

   c. intolerance to heat.
   d. apathy.

18. Which of the following is often responsible for low iron intakes in the western world?

    a. blood loss caused by parasitic infections of the digestive tract
    b. malnutrition which includes an inadequate iron intake
    c. vegetarians who design their own meal patterns
    d. displacement of nutrient-rich foods by foods high in sugar and fat

19. The usual source of fluoride is:

    a. drinking water.
    b. cooking utensils.
    c. fortified foods.
    d. processed foods.

20. People with severe iodine deficiency:

    a. become sluggish.
    b. develop a goiter.
    c. lose weight.
    d. a and b
    e. b and c

21. Temporary fluctuations in body water

    a. reflect a change in body fat.
    b. show up on the scale.
    c. reflect gain or loss in water weight.
    d. a and b
    e. b and c

22. Given a normal diet and moderate environmental conditions, women need about _____ cups of fluid from drinking water and beverages per day.

    a. 3
    b. 5
    c. 9
    d. 13

23. Which of the following is the best source of phosphorus?

    a. steak
    b. beans
    c. bread
    d. cabbage

24. The DASH diet calls for greatly increased intakes of _____.

    a. nuts
    b. whole grains
    c. fish
    d. fruits and vegetables

Application Level Items:

25. Which of the following directly determines whether calcium is withdrawn from or deposited to the skeleton?

    a. hormones that are sensitive to blood levels of calcium
    b. the amount of calcium consumed in the diet
    c. the amount of calcium deposited in bones during peak bone mass
    d. a and b
    e. b and c

26. Why is the intake recommendation for calcium set at 1300 milligrams/day for adolescents?

    a. because the skeleton begins to lose bone density during that age
    b. because people develop their peak bone mass during this time
    c. because a high calcium intake is thought to maximize bone density during the growing years
    d. a and b
    e. b and c

27. Which of the following provides the best choices for high-calcium food selections?

    a. ice milk, cottage cheese, butter, and cream
    b. milk, buttermilk, cheese, and yogurt
    c. cream cheese, chocolate milk, and frozen yogurt
    d. sardines, spinach, kale, and Swiss chard

28. How much sodium is consumed by someone who eats five grams of salt?

    a. 1000 mg
    b. 1500 mg
    c. 2000 mg
    d. 2500 mg

29. An increase in iodine intake in the United States has resulted from:

    a. dough conditioners used in the baking industry.
    b. milk produced in dairies that feed cows iodine-containing medications.
    c. the increased consumption of fast foods and convenience items.
    d. a and b
    e. a and c

30. To improve the absorption of iron from an iron supplement you would:

    a. take the supplement with orange juice.
    b. take the supplement with tea.
    c. consume a small amount of meat along with the supplement.
    d. a and b
    e. a and c

Answers

Summing Up

1. 60
2. satiety
3. thirst
4. kidneys
5. brain
6. weight
7. balance
8. transportation
9. temperature
10. 13
11. 80
12. magnesium
13. sodium
14. cells
15. outsides
16. minerals
17. electrolytes
18. fluid
19. acid-base
20. bases
21. abundant
22. tooth
23. nerve
24. osteoporosis
25. peak bone mass
26. outside
27. 40
28. acid-base
29. 1500
30. diuretics
31. sulfate
32. chloride
33. thyroxine
34. iron
35. myoglobin
36. enzymes
37. teeth

Chapter Glossary

Matching Exercise:

1. e	5. r	9. a	13. l	17. c
2. j	6. d	10. q	14. f	18. n
3. g	7. o	11. m	15. p	19. t
4. b	8. k	12. h	16. i	20. s

Crossword Puzzle:

1. major minerals
2. water intoxication
3. fluorapatite
4. electrolytes
5. anemia
6. minerals
7. diuretics
8. water balance
9. trace minerals
10. fluorosis
11. leavened
12. soft water (across); salts (down)
13. ions
14. hypertension
15. hard water

Exercises

Chapter Study Questions:

1. Hard water has high concentrations of calcium and magnesium, while soft water's principal mineral is sodium. Soft water adds appreciable amounts of sodium to the diet and may aggravate hypertension and heart disease. Hard water may oppose these conditions because of its calcium content. Soft water also dissolves cadmium and lead from pipes. Cadmium is suspected of promoting bone fractures, kidney problems, and hypertension and lead poisoning is especially harmful to children.

2. Because absorption of calcium varies not only with age, but also with a person's vitamin D status and the calcium content of the diet.

3. Because individuals develop their peak bone mass, which is the highest attainable bone density for an individual, during childhood and adolescence. Around age 30, the skeleton no longer adds significantly to bone density and the bones begin to lose density.

4. Diuretics are medications which cause increased water excretion and dehydration which leads to potassium loss from inside cells. It is especially dangerous because potassium loss from brain cells makes the victim unaware of the need for water. Potassium-rich foods are necessary to compensate for the losses.

5. Pica is the consumption of non-food substances, such as ice, clay, paste, or soil, and is associated with iron deficiency.

6. DASH stands for Dietary Approaches to Stop Hypertension and it helps to lower blood pressure. The diet emphasizes increased intakes of fruits and vegetables and adequate intakes of nuts, fish, whole grains, and low-fat dairy products. Red meat, butter, other high-fat foods, sweets and salt and sodium are reduced.

Short Answer Questions:

1. (a) calcium; (b) chloride; (c) magnesium; (d) phosphorus; (e) potassium; (f) sodium; (g) sulfate

2. (a) calcium; (b) magnesium; (c) sodium

3. (a) regulates transport of ions across cell membranes and is important in nerve transmission; (b) helps maintain normal blood pressure; (c) is essential for muscle contraction and maintenance of heartbeat; (d) plays a role in clotting of blood; (e) allows secretion of hormones, digestive enzymes, and neurotransmitters; (f) activates cellular enzymes that regulate many processes

4. (a) overly strict use of low-sodium diets; (b) vomiting; (c) diarrhea; (d) heavy sweating

5. (a) learn to control use of the salt shaker; (b) limit intake of processed and fast foods

6. (a) inadequate intake; (b) vomiting; (c) diarrhea; (d) alcoholism; (e) protein malnutrition

7. (a) seafood; (b) iodized salt

8. (a) tannins of tea and coffee; (b) calcium and phosphorus in milk; (c) phytates that accompany fiber in lightly processed whole-grain cereals and legumes

Problem-Solving:

Decrease the amount of salt; substitute ¼ cup oil for the ¹/₃ cup lard; use fresh or frozen green peas rather than canned green peas; omit the ¼ cup butter; omit the poultry seasoning; use unsalted and fat-free chicken broth; substitute skim milk for the cream.

Controversy Questions:

1. Trabecular bone is the web-like structure composed of calcium-containing crystals inside a bone's solid outer shell; it provides strength and acts as a calcium storage bank. Cortical bone is the dense, ivory-like bone that forms the exterior shell of a bone and the shaft of a long bone; it provides a sturdy outer wall.

2. Poor nutrition involving calcium and vitamin D; estrogen deficiency in women; lack of physical activity; being underweight; use of alcohol and tobacco; and excess protein, sodium, caffeine, and soft drinks and inadequate vitamin K.

3. It is thought that genetic inheritance influences the maximum bone mass possible during growth and the extent of a woman's bone loss during menopause.

4. Weight-bearing exercises such as calisthenics, dancing, jogging, vigorous walking or weight training.

5. Using milk as the primary beverage and consuming a balanced diet that provides all nutrients; playing actively or engaging in sports; limiting television and other sedentary activities; drinking fluoridated water; and not starting to smoke or drink alcohol.

Study Aids

Major Minerals			
Name	Chief Functions	Deficiency Disease Name	Food Sources
Calcium	• makes up bone and <u>teeth</u> structure • normal muscle contraction • nerve functioning • blood <u>clotting</u> • immune defenses • blood pressure	<u>osteoporosis</u>	• milk and milk products • small <u>fish</u> with bones • tofu • certain leafy greens • legumes

Major Minerals			
Name	Chief Functions	Deficiency Disease Name	Food Sources
Magnesium	• bone mineralization • building of protein • normal muscle contraction • transmission of nerve impulses • maintenance of teeth		• nuts • legumes • whole grains • dark green vegetables • seafoods • chocolate • cocoa
Sodium	• maintains normal fluid balance • maintains normal acid-base balance		• salt • processed foods • soy sauce

Trace Minerals			
Name	Chief Functions	Deficiency Disease Name	Food Sources
Iron	• part of the protein hemoglobin • part of the protein myoglobin • necessary for utilization of energy in the body	anemia	• red meats • fish • poultry • shellfish • eggs • legumes • dried fruits
Iodine	• component of thyroid hormone thyroxine	goiter cretinism	• seafood • iodized salt • bread
Zinc	• part of insulin and many enzymes • normal fetal development		• protein-containing foods

Sample Test Items

1. c (p. 270)
2. d (p. 273)
3. a (p. 275)
4. d (p. 273)
5. c (p. 280)
6. d (p. 282)
7. d (p. 282)
8. a (p. 282)
9. d (p. 312)
10. c (p. 287)
11. d (p. 288)
12. c (p. 288)
13. d (pp. 289-290)
14. a (p. 291)
15. e (pp. 285-286)
16. b (p. 292)
17. c (p. 294)
18. d (p. 295)
19. a (p. 301)
20. d (p. 292)
21. e (p. 272)
22. c (pp. 273-274)
23. a (p. 284)
24. d (p. 287)
25. a (pp. 281-282)
26. e (p. 283)
27. b (p. 306)
28. c (p. 286)
29. e (p. 293)
30. e (p. 298)

Chapter 9 - Energy Balance and Healthy Body Weight

Chapter Objectives

After completing this chapter, you should be able to:

1. Discuss the problems of too much or too little body fat.

2. Describe health risks associated with central obesity and those groups most likely to be affected.

3. List and define the major components of the body's energy budget.

4. Identify and explain the factors that affect the basal metabolic rate.

5. Estimate an individual's total energy requirement.

6. Discuss the role of BMI in evaluating health risks of obesity.

7. Evaluate the methods used for estimating body fatness.

8. Summarize the theories that attempt to explain the mystery of obesity.

9. Explain what happens during moderate weight loss versus rapid weight loss.

10. Identify the pros and cons of high-protein, low-carbohydrate diets.

11. Explain strategies that work best for weight gain.

12. Summarize the recommended diet strategies to promote weight loss and maintenance.

13. Describe how physical activity contributes to a weight loss program.

14. Define and describe eating disorders and explain the physical harm that occurs as a result of these behaviors (Controversy 9).

Key Concepts

✓ Deficient body fatness threatens survival during a famine or during diseases.

✓ Most obese people suffer illnesses, and obesity is considered a chronic disease.

✓ Central obesity may be more hazardous to health than other forms of obesity.

✓ Experts estimate health risks from obesity using BMI, waist circumference, and a disease risk profile. Fit people are healthier than unfit people of the same body fatness.

✓ Overfatness presents social and economic handicaps as well as physical ills. Judging people by their body weight is a form of prejudice in our society.

✓ The "energy in" side of the body's energy budget is measured in calories taken in each day in the form of foods and beverages. The number of calories in foods and beverages can be obtained from published tables or computer diet analysis programs. No easy method exists for determining the "energy out" side of a person's energy balance equation.

✓ Two major components of the "energy out" side of the body's energy budget are basal metabolism and voluntary activities. A third component of energy expenditure is the thermic effect of food. Many factors influence the basal metabolic rate.

- The DRI Committee sets Estimated Energy Requirements for a reference man and woman. People's energy needs vary greatly.

- The DRI Committee has established a method for determining an individual's approximate energy requirement.

- The BMI values mathematically correlate heights and weights with risks to health. They are especially useful for evaluating health risks of obesity but fail to measure body composition or fat distribution.

- A clinician can determine the percentage of fat in a person's body by measuring fatfolds, body density, or other parameters. Distribution of fat can be estimated by radiographic techniques, and central adiposity can be assessed by measuring waist circumference.

- No single body composition or weight suits everyone; needs vary by gender, lifestyle, and stage of life.

- Hunger is stimulated by an absence of food in the digestive tract. Appetite can occur with or without hunger, and many factors affect it.

- Satiation occurs when the digestive organs signal the brain that enough food has been consumed. Satiety is the feeling of fullness that lasts until the next meal. Hunger outweighs satiety in the appetite control system.

- The adipose tissue hormone leptin suppresses the appetite in response to a gain in body fat.

- Some foods may confer greater satiety than others, but these effects are not yet established scientifically.

- Metabolic theories attempt to explain obesity on the basis of molecular functioning. Quacks often exploit these theories for profit.

- A person's genetic inheritance greatly influences, but does not ensure, the development of obesity.

- Studies of human behavior identify stimuli that lead to overeating. Food pricing, availability, and advertising influence food choices. Physical inactivity is clearly linked with overfatness.

- When energy balance is negative, glycogen returns glucose to the blood. When glycogen runs out, body protein is called upon for glucose. Fat also supplies fuel as fatty acids. If glucose runs out, fat supplies fuel as ketone bodies, but ketosis can be dangerous. Both prolonged fasts and low-carbohydrate diets are ill-advised.

- When energy balance is positive, carbohydrate is converted to glycogen or fat, protein is converted to fat, and food fat is stored as fat. Alcohol delivers calories and encourages fat storage.

- To achieve and maintain a healthy body weight, set realistic goals, keep records, and expect to progress slowly. Watch energy density, make the diet adequate and balanced, limit calories, reduce alcohol, and eat regularly, especially at breakfast. Night eating can be a problem.

- Physical activity greatly augments diet in weight-loss efforts. Improvements in health and body composition follow an active lifestyle.

- Weight gain requires a diet of calorie-dense foods, eaten frequently throughout the day. Physical activity builds lean tissue, and no special supplements can speed the process.

- For people whose obesity threatens their health, medical science offers drugs and surgery. The effectiveness of herbal products and other gimmicks has not been demonstrated, and they may prove hazardous.

140

✓ People who succeed at maintaining lost weight keep to their eating routines, keep exercising, and keep track of calorie and fat intakes and body weight. The more traits related to positive self-image and self-efficacy a person possesses or cultivates, the more likely that person will succeed.

Summing Up

Both deficient and (1)_____ body fat present health risks. For example, underweight increases the risk for any person fighting a (2)_____ disease, while overfatness can precipitate (3)_____ and thus increase the risk of stroke. Being overfat also triples a person's risk of developing (4)_____. The (5)_____ usually correlates with degree of body fatness and disease risks.

The "energy in" side of the body's energy budget is measured in (6)_____ taken in each day in the form of foods and beverages. Calories in foods and beverages can be obtained from published tables or (7)_____ programs. The body spends energy in two major ways: to fuel its (8)_____ metabolism and to fuel its (9)_____ activities.

The DRI Committee sets (10)_____ values by taking into account the ways in which energy is spent and by whom. A clinician can determine the percentage of fat in a person's body by measuring skinfolds, body (11)_____, or other parameters. Distribution of fat can be estimated by radiographic techniques and central adiposity by measuring (12)_____ circumference.

Different theories exist to explain the development of obesity, with one attributing it to internal factors and the other to (13)_____ factors. The influence of (14)_____ on obesity has been demonstrated in children, whose body fatness often resembles that of their parents. A different line of research contends that obesity is determined by behavioral responses to environmental stimuli, which forms the rationale for the importance of (15)_____ cues to overeating. Physical (16)_____ is clearly linked with overfatness.

Researchers are interested in why people eat when and what they eat, and especially why some people overeat, and this has lead to investigations of hunger, (17)_____, and satiety. Hunger is the (18)_____ need to eat, while appetite is the (19)_____ desire to eat. The perception of (20)_____ that lingers in the hours after a meal is called satiety. (21)_____ is a hormone produced by the adipose tissue that is directly linked to both appetite control and body fatness. Of the three energy-yielding nutrients, (22)_____ is perhaps the most satiating.

Changes in body weight can reflect shifts in body fluid content, in bone minerals, or in (23)_____ tissues such as muscles. The type of (24)_____ gained or lost depends on how the person goes about gaining or losing it. With prolonged fasting or carbohydrate deprivation, the body adapts by converting (25)_____ into compounds that the nervous system can adapt to use and this condition is called (26)_____.

Many different approaches to weight loss are used including fasting and low-(27)_____ diets which are ill-advised. People who successfully lose and maintain weight loss set realistic goals, keep (28)_____, and expect to progress slowly.

Chapter Glossary

Matching Exercise:

_____ 1. leptin

_____ 2. ghrelin

_____ 3. thermogenesis

_____ 4. wasting

_____ 5. appetite

_____ 6. thermic effect of food

_____ 7. satiation

_____ 8. behavior therapy

_____ 9. adipose tissue

_____ 10. ketone bodies

_____ 11. body composition

_____ 12. extreme obesity

_____ 13. lipoprotein lipase

_____ 14. hunger

_____ 15. bioelectrical impedance

_____ 16. energy density

_____ 17. overweight

_____ 18. underweight

a. the body's speeded-up metabolism in response to having eaten a meal

b. acidic compounds derived from fat and certain amino acids

c. a technique to measure body fatness by measuring the body's degree of electrical conductivity

d. an appetite-suppressing hormone produced in the fat cells that conveys information about body fatness to the brain

e. alteration of behavior using methods based on the theory that actions can be controlled by manipulating the environmental factors that cue, or trigger, the actions

f. an enzyme mounted on the surfaces of fat cells that splits triglycerides in the blood into fatty acids and glycerol to be absorbed into the cells for reassembly and storage

g. the generation and release of body heat associated with the breakdown of body fuels

h. the body's fat tissue, which performs several functions

i. a hormone released by the stomach that signals the hypothalamus to stimulate eating

j. the psychological desire to eat; a learned motivation and a positive sensation that accompanies the sight, smell, or thought of appealing foods

k. the proportions of muscle, bone, fat, and other tissue that make up a person's total body weight

l. a measure of the energy provided by a food relative to its weight

m. the physiological need to eat, experienced as a drive for obtaining food; an unpleasant sensation that demands relief

n. the progressive, relentless loss of the body's tissues that accompanies certain diseases and shortens survival time

o. the perception of fullness that builds throughout a meal, eventually reaching the degree of fullness and satisfaction that halts eating

p. the condition of having a body mass index of 40 or above

q. having a body mass index of less than 18.5

r. having a body mass index of 25.0 – 29.9

Crossword Puzzle:

Across:

3. measurement of the thickness of a fold of skin on the back of the arm (over the triceps muscle), below the shoulder blade (subscapular), or in other places, using a caliper
5. fat stored within the abdominal cavity in association with the internal abdominal organs
6. intentional activities conducted by voluntary muscles
11. the rate at which the body uses energy to support its basal metabolism
12. a measure of density and volume used to determine body fat content
14. overfatness, with adverse health effects
15. excess fat in the abdomen and around the trunk

Down:

1. the sum total of all the involuntary activities that are necessary to sustain life, including respiration, circulation, and new tissue synthesis, but excluding digestion and voluntary activities
2. an indicator of obesity, calculated by dividing the weight of a person by the square of the person's height
4. a term popularly used to describe dimpled fat tissue on the thighs and buttocks; not recognized in science
7. compounds of the brain whose actions mimic those of opiate drugs in reducing pain and producing pleasure
8. fat stored directly under the skin
9. a person's belief in his or her ability to succeed in an undertaking
10. repeated rounds of weight loss and subsequent regain, with reduced ability to lose weight with each attempt
13. a noninvasive method of determining total body fat, fat distribution and bone density

Exercises

Answer these chapter study questions:

1. What are risks associated with being underweight?

2. What are two major drawbacks of BMI values?

3. Identify appropriate strategies for weight gain.

4. Describe diet-related changes that most often lead to successful weight change and maintenance.

5. Describe how exercise can increase the BMR.

6. What factors are considered by the DRI Committee in setting Estimated Energy Requirements (EER)?

Complete these short answer questions:

1. Obesity elevates the risk of these major conditions:

    a.   c.

    b.   d.

2. Two major components of the "energy out" side of the body's energy budget are:

    a.   b.

3. Body mass index is of limited use with:

    a.   c.

    b.

4. Techniques for estimating body fatness include:

   a.   c.

   b.   d.

5. Leptin is a hormone produced by adipose tissue that is directly linked to:

   a.   b.

6. Gradual weight loss is preferred to rapid weight loss because:

   a.

   b.

7. Changes in body weight may reflect shifts in any of the following materials:

   a.

   b.

   c.

   d.

8. The three kinds of energy nutrients are stored in the body in two forms, including

   _____ and _____ .

9. Benefits of physical activity in a weight-management program include:

   a.

   b.

   c.

   d.

   e.

   f.

   g.

10. In late food deprivation, _____ bodies help feed the nervous system and so help spare tissue protein.

Solve these problems:

1. Use the quick and easy estimate of energy needs on page 326 of your textbook to calculate the number of calories needed each day by a 110-pound female who is 5' 3" tall and is sedentary. She is 25 years old.

2. Use the concept of energy density to modify the following breakfast menu.

Menu   **Changes**

Omelet made with cheese   _____

                             _____

White toast with butter   _____

                             _____

Hash brown potatoes   _____

                             _____

Whole milk   _____

                             _____

Apple juice   _____

                             _____

3. Calculate the grams of carbohydrate, protein, and fat which should be supplied in a 1000-calorie diet, using the distribution of 50% carbohydrate, 25% protein, and 25% fat.

Answer these controversy questions:

1. What are the three associated medical problems that form the female athlete triad?

2. What are three characteristics specific to anorexia nervosa?

3. What two issues and behaviors must be addressed in the treatment of anorexia nervosa?

4. What are characteristics of typical binge foods that are consumed during a bulimic binge?

5. What are the two primary goals of treatment for bulimia nervosa?

Study Aids

1. Identify whether the following factors are associated with a higher or lower basal metabolic rate (BMR) by placing an X in the appropriate space.

Factor	Higher BMR	Lower BMR
a. Youth	_____	_____
b. Being tall and thin	_____	_____
c. Starvation	_____	_____
d. Heat	_____	_____
e. Fasting	_____	_____
f. Stress hormones	_____	_____
g. Pregnant women	_____	_____
h. Fever	_____	_____
i. Thyroxine	_____	_____

2. Fill in the missing terms in this diagram on feasting and fasting.

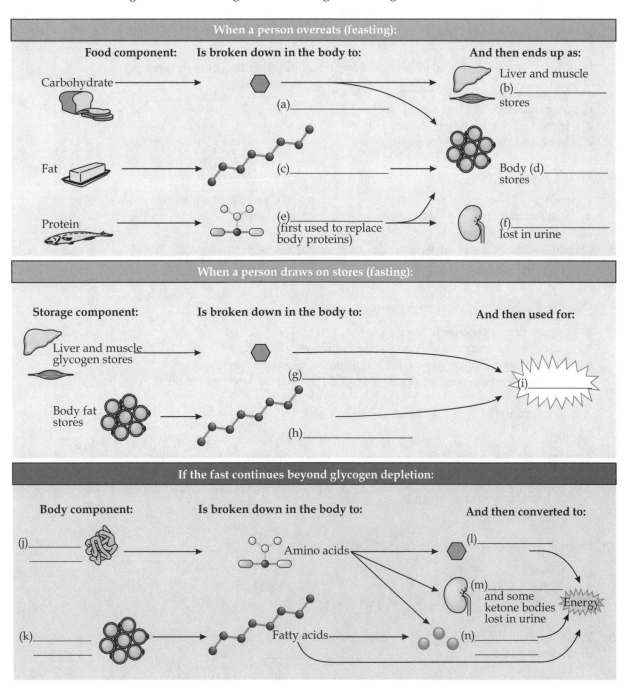

When a person overeats (feasting):

Food component: Is broken down in the body to: And then ends up as:

Carbohydrate

(a)_____

Liver and muscle
(b)_____
stores

Fat

(c)_____

Body (d)_____
stores

Protein

(e)_____
(first used to replace
body proteins)

(f)_____
lost in urine

When a person draws on stores (fasting):

Storage component: Is broken down in the body to: And then used for:

Liver and muscle
glycogen stores

(g)_____

Body fat
stores

(h)_____

(i)_____

If the fast continues beyond glycogen depletion:

Body component: Is broken down in the body to: And then converted to:

(j)_____

Amino acids

(l)_____

(m)_____
and some
ketone bodies
lost in urine

(k)_____

Fatty acids

(n)_____

Energy

149

Comprehension Level Items:

1. Health risks associated with excessive body fat include all of the following **except**:

   a. increased risk of developing hypertension.
   b. increased risk for developing ulcers.
   c. increased risk for heart disease.
   d. increased risk for developing diabetes.

2. Which of the following groups is **least** likely to carry more intraabdominal fat?

   a. men
   b. women in their reproductive years
   c. women past menopause
   d. smokers

3. Obesity is defined as a body mass index of 30 or higher.

   a. true
   b. false

4. Outside-the-body causes of obesity suggest that:

   a. people eat in response to internal factors.
   b. people eat in response to factors such as hunger.
   c. environmental influences override internal regulatory systems.
   d. the body tends to maintain a certain weight by means of external controls.

5. A man of normal weight may have, on the average, _____ percent of the body weight as fat.

   a. 5-10
   b. 10-15
   c. 12-20
   d. 20-30

6. BMI values are most valuable for evaluating health risks associated with underweight or overweight.

   a. true
   b. false

7. Central obesity correlates with an increased incidence of:

   a. diabetes.
   b. kidney disease.
   c. stroke.
   d. a and b
   e. a and c

8. Which of the following may help to explain the association seen between overweight and sleep deprivation?

   a. satiety
   b. leptin
   c. night eating syndrome
   d. endorphins

9. Which of the following helps to feed the brain during times when too little carbohydrate is available?

   a. lipoprotein lipase
   b. brown fat
   c. ketone bodies
   d. leptin

10. Natural herbs are never harmful to the body.

    a. true

    b. false

11. Which of the following largely determines whether weight is gained as body fat or as lean tissue?

    a. the number of calories consumed
    b. the amount of fat consumed
    c. the person's current body composition
    d. exercise

12. The between-meal interval is normally _____ waking hours.

    a. 2 – 4
    b. 4 – 6
    c. 8 – 12
    d. 12 – 14

13. Ketone bodies are:

    a. acidic compounds.
    b. derived from fat.
    c. alkaline compounds.
    d. a and b
    e. b and c

14. Which of the following directly controls basal metabolism?

    a. epinephrine
    b. insulin
    c. thyroxine
    d. lipase

15. Which of the following statements is **not** true?

    a. Any food can make you fat if you eat enough of it.
    b. Alcohol both delivers calories and encourages fat accumulation.
    c. Fat from food, as opposed to carbohydrate or protein, is especially easy for the body to store as fat tissue.
    d. Protein, once converted to fat, can later be recovered as amino acids.

16. Diet strategies for weight loss should include all of the following **except**:

    a. skipping meals.
    b. using the concept of energy density.
    c. choosing realistic calorie levels.
    d. keeping records.

17. Skinfold measurements provide an accurate estimate of total body fat and a fair assessment of the fat's location.

    a. true

    b. false

18. Which of the following is the first step in making diet-related changes that most often lead to successful weight loss and maintenance?

    a. Reward yourself personally and immediately.
    b. Strengthen cues to appropriate behaviors.
    c. Set appropriate goals.
    d. Arrange negative consequences for negative behaviors.

19. All of the following are considered to be part of basal metabolism **except**:

   a. beating of the heart.
   b. inhaling and exhaling of air.
   c. maintenance of body temperature.
   d. sitting and walking.

20. Fat is the most satiating of the three energy-yielding nutrients.

   a. true
   b. false

21. Gradual weight loss is preferred to rapid weight loss because:

   a. lean body mass is spared.
   b. fat is lost.
   c. lean body mass is lost.
   d. a and b
   e. b and c

22. Modest weight loss can lead to rapid improvements in control over diabetes, blood pressure, and blood lipids.

   a. true
   b. false

23. According to data from the National Weight Control Registry, people who are successful at weight loss and weight maintenance:

   a. change their food intakes between weekdays and weekends.
   b. eat breakfast.
   c. eat high-fiber foods.
   d. a and b
   e. b and c

24. The DRI Committee recommends at least _____ grams of carbohydrates a day.

   a. 120
   b. 125
   c. 130
   d. 135

Application Level Items:

25. Kathy weighs 140 pounds and has a body mass index of 28. Based on this information you would classify her as:

   a. underweight.
   b. normal weight.
   c. overweight.
   d. obese.

26. John has a body mass index of 31; therefore, you would classify him as:

   a. underweight.
   b. normal weight.
   c. overweight.
   d. obese, class I.

27. Susie is underweight and wishes to gain weight. Which of the following should she do?

   a. exercise
   b. eat only three meals a day
   c. eat a high-calorie diet
   d. a and b
   e. a and c

28. Emily has been overweight most of her life and she has been on almost every type of diet possible. Although she loses weight in the first few months of following the diet, a year later she starts to regain. Which of the following types of diets are most consistent with Emily's experiences?

    a. low-fat diet   c. low-protein diet
    b. low-carbohydrate diet   d. low-calorie diet

29. Sarah is bragging about losing 10 pounds in one week on a weight-loss program. What would you tell Sarah?

    a. It's great that you have lost that much weight so fast.
    b. You need to slow down your rate of weight loss.
    c. The weight loss probably represents changes in body fluid content.
    d. Your change in weight probably reflects a loss of body fat.

30. Karen is interested in speeding up her basal metabolic rate to promote fat loss. You would advise her to:

    a. eat smaller meals more frequently throughout the day.
    b. participate in endurance and strength-building exercise daily.
    c. fast at least one day each week.
    d. follow a low-carbohydrate diet.

Answers

Summing Up

1. excessive
2. wasting
3. hypertension
4. diabetes
5. body mass index
6. calories
7. computer diet analysis
8. basal
9. voluntary
10. Estimated Energy Requirement
11. density
12. waist
13. external
14. genetics
15. external
16. inactivity
17. appetite
18. physiological
19. psychological
20. fullness
21. leptin
22. protein
23. lean
24. tissue
25. fat
26. ketosis
27. carbohydrate
28. records

Chapter Glossary

Matching Exercise:

1. d
2. i
3. g
4. n
5. j
6. a
7. o
8. e
9. h
10. b
11. k
12. p
13. f
14. m
15. c
16. l
17. r
18. q

Crossword Puzzle:

1. basal metabolism
2. body mass index
3. skinfold test
4. cellulite
5. visceral fat
6. voluntary activities
7. endorphins
8. subcutaneous fat
9. self-efficacy
10. weight cycling
11. basal metabolic rate
12. underwater weighing
13. DEXA
14. obesity
15. central obesity

Exercises

Chapter Study Questions:

1. Overly thin people are at a disadvantage in the hospital where they may have to refrain from eating food for days at a time so that they can undergo tests or surgery and their nutrient status can easily deteriorate. Underweight also increases the risk for any person fighting a wasting disease such as cancer.

2. They fail to indicate how much of the weight is fat and where the fat is located.

3. Participate in physical activity, especially strength training; increase calories and learn to eat nutritious, calorie-dense foods; eat more frequently; increase portion sizes; keep ready-to-eat foods on hand for quick meals.

4. Keep records of food intake and exercise habits; set realistic goals; plan a diet with foods you like; pay attention to portions; eat regularly; include breakfast; cut down on or eliminate alcohol; watch energy density; make the diet adequate and balanced with carbohydrates, protein, and fat; limit calories.

5. Exercise helps body composition change toward the lean. Lean tissue is more metabolically active than fat tissue, so the basal energy output picks up the pace as well.

6. A person's gender, age, physical activity and body size and weight are considered in estimating a person's EER. Also the BMR is high in people who are growing, so pregnant women and children have their own sets of energy equations.

Short Answer Questions:

1. (a) hypertension; (b) heart disease; (c) stroke; (d) diabetes

2. (a) basal metabolism; (b) voluntary activities

3. (a) athletes; (b) adults over 65; (c) pregnant and lactating women

4. (a) anthropometry; (b) density; (c) conductivity; (d) radiographic techniques

5. (a) appetite control; (b) body fatness

6. (a) lean body mass is spared; (b) fat is lost

7. (a) fluid content; (b) bone minerals; (c) lean tissue; (d) contents of the bladder or digestive tract

8. glycogen; fat

9. (a) short-term increase in energy expenditure; (b) long-term increase in BMR; (c) improved body composition; (d) appetite control; (e) stress reduction and control of stress eating; (f) physical and psychological well-being; (g) improved self-esteem

10. ketone

Problem-Solving:

1. (a) change pounds to kilograms: 110 pounds divided by 2.2. pounds = 50 kilograms;
   (b) multiply kilograms body weight by 22 = calories/day; 50 kilograms x 22 = 1100 calories/day

2. **Menu** — **Changes**

Menu	Changes
Omelet made with cheese	Substitute a boiled egg, which is lower in fat, or add vegetables to a regular omelet to decrease fat and increase fiber
White toast with butter	Substitute whole-wheat toast with a small amount of margarine to increase fiber and decrease fat
Hash brown potatoes	Eliminate hash brown potatoes to decrease fat
Whole milk	Substitute skim milk to decrease fat
Apple juice	Substitute sliced fresh apple for apple juice to increase fiber and water

3. (a) 50% carbohydrate × 1000 calories = 500 carbohydrate calories divided by 4 calories per gram = 125 grams carbohydrate;

   (b) 25% protein × 1000 calories = 250 protein calories divided by 4 calories per gram = 63 grams protein;

   (c) 25% fat × 1000 calories = 250 fat calories divided by 9 calories per gram = 28 grams fat

Controversy Questions:

1. Disordered eating, amenorrhea, and osteoporosis.

2. A refusal to maintain a body weight at or above a minimal normal weight for age and height; intense fear of gaining weight or becoming fat; undue influence of body weight or shape on self-evaluation.

3. Those relating to food and weight and those involving relationships with oneself and others.

4. Typical foods are easy-to-eat, low-fiber, smooth-textured, high-fat, and high-carbohydrate foods.

5. Steady maintenance of weight and prevention of relapse into cyclic gains and losses.

Study Aids

1.

Factor	Higher BMR	Lower BMR
a. Youth	X	
b. Being tall and thin	X	
c. Starvation		X
d. Heat	X	
e. Fasting		X
f. Stress hormones	X	
g. Pregnant women	X	
h. Fever	X	
i. Thyroxine	X	

2. (a) glucose; (b) glycogen; (c) fatty acids; (d) fat; (e) amino acids; (f) nitrogen; (g) glucose; (h) fatty acids; (i) energy; (j) body protein; (k) body fat; (l) glucose; (m) nitrogen; (n) ketone bodies

Sample Test Items

1. b (p. 321)
2. b (pp. 321-322)
3. a (p. 327)
4. c (pp. 335-336)
5. c (p. 329)
6. a (p. 326)
7. e (p. 321)
8. b (p. 332)

9. c (p. 338)
10. b (p. 354)
11. d (p. 340)
12. b (p. 338)
13. d (p. 338)
14. c (p. 325)
15. d (p. 340)
16. a (pp. 344-348)

17. a (p. 327)
18. c (p. 344)
19. d (p. 324)
20. b (p. 332)
21. d (p. 338)
22. a (p. 343)
23. e (p. 355)
24. c (p. 340)

25. c (p. 322)
26. d (p. 322)
27. e (p. 351)
28. b (p. 341)
29. c (p. 337)
30. b (p. 325)

Chapter 10 - Nutrients, Physical Activity, and the Body's Responses

Chapter Objectives

After completing this chapter, you should be able to:

1. Explain the benefits of and guidelines for regular physical activity for the body.

2. Describe the components of fitness and how muscles respond to physical activity.

3. Summarize how the body adjusts its fuel mix to respond to physical activity of varying intensity levels and duration.

4. Explain the concept of carbohydrate loading.

5. Discuss the roles vitamins and minerals play in physical performance and indicate whether supplements are necessary to support the needs of active people.

6. Describe the importance of fluids and temperature regulation in physical activity, and the best way to stay hydrated before and during exercise.

7. Discuss conditions under which athletes benefit from sports drinks.

8. Describe the characteristics of and plan an appropriate diet for athletes.

9. List the risks of taking "ergogenic" aids and steroids to enhance physical performance (Controversy 10).

Key Concepts

✓ Physical activity and fitness benefit people's physical and psychological well-being and improve their resistance to disease. Physical activity to improve physical fitness offers additional personal benefits.

✓ The components of fitness are flexibility, muscle strength, muscle endurance, and cardiorespiratory endurance. To build fitness, a person must engage in physical activity. Muscles adapt to activities they are called upon to perform repeatedly.

✓ Weight training offers health and fitness benefits to adults. Weight training reduces the risk of cardiovascular disease, improves older adults' physical mobility, and helps maximize and maintain bone mass.

✓ Cardiorespiratory endurance training enhances the ability of the heart and lungs to deliver oxygen to the muscles. With cardiorespiratory endurance training, the heart becomes stronger, breathing becomes more efficient, and the health of the entire body improves.

✓ Glucose is supplied by dietary carbohydrate or made by the liver. It is stored in both liver and muscle tissue as glycogen. Total glycogen stores affect an athlete's endurance.

✓ The more intense an activity, the more glucose it demands. During anaerobic metabolism, the body spends glucose rapidly and accumulates lactate.

✓ Physical activity of long duration places demands on the body's glycogen stores. Carbohydrate ingested before and during long-duration activity may help to forestall hypoglycemia and fatigue. Carbohydrate loading is a regimen of physical activity and diet that enables an athlete's muscles to store larger-than-normal amounts of glycogen to extend endurance. After strenuous training, eating foods with a high glycemic index may help restore glycogen most rapidly.

✓ Highly trained muscles use less glucose and more fat than do untrained muscles to perform the same work, so their glycogen lasts longer.

✓ Athletes who eat high-fat diets may burn more fat during endurance activity, but the risks to health outweigh any possible performance benefits. The intensity and duration of activity, as well as the degree of training, affect fat use.

✓ Physical activity stimulates muscle cells to break down and synthesize protein, resulting in muscle adaptation to activity. Athletes use protein both for building muscle tissue and for energy. Diet, intensity and duration of activity, and training affect protein use during activity.

✓ Although athletes need more protein than sedentary people, a balanced, high-carbohydrate diet provides sufficient protein to cover an athlete's needs.

✓ Vitamins are essential for releasing the energy trapped in energy-yielding nutrients and for other functions that support physical activity. Active people can meet their vitamin needs if they eat enough nutrient-dense foods to meet their energy needs. Iron-deficiency anemia impairs physical performance because iron is the blood's oxygen handler. Sports anemia is a harmless temporary adaptation to physical activity.

✓ Evaporation of sweat cools the body. Heat stroke can be a threat to physically active people in hot, humid weather. Hypothermia threatens those who exercise in the cold.

✓ Physically active people lose fluids and must replace them to avoid dehydration. Thirst indicates that water loss has already occurred.

✓ Water is the best drink for most physically active people, but endurance athletes need drinks that supply glucose as well as fluids.

✓ The body adapts to compensate for sweat losses of electrolytes. Athletes are advised to use foods, not supplements, to make up for these losses.

✓ During events lasting longer than three hours, athletes need to pay special attention to replacing sodium losses to prevent hyponatremia.

✓ Caffeine-containing drinks within limits may not impair performance, but water and fruit juice are preferred. Alcohol use can impair performance in many ways and is not recommended.

Summing Up

Physical activity benefits people's physical and psychological well-being and improves their resistance to (1)_____. Fitness is composed of four components including flexibility, muscle (2)_____, (3)_____ endurance and cardiorespiratory endurance. Muscle cells and tissues respond to an (4)_____ of physical activity by gaining strength and size, a response called (5)_____. The kind of exercise that brings about cardiorespiratory endurance is (6)_____ activity. Weight training builds

(7)_____ and develops and maintains (8)_____ strength and endurance.

In the early minutes of an activity, muscle (9)_____ provides the majority of energy the muscles use to go into action. How long an exercising person's glycogen will last depends not only on diet but also partly on the (10)_____ of the activity. Glycogen depletion usually occurs after about (11)_____ hours of vigorous exercise. Glucose use during physical activity depends not only on the intensity, but also on the (12)_____ of the activity. A person who continues to exercise moderately for longer than 20 minutes begins to use less glucose and more (13)_____ for fuel. One diet strategy useful for endurance athletes to help maintain glucose concentration is to eat a high-(14)_____ diet on a day-to-day basis.

Many vitamins and minerals assist in releasing (15)_____ from fuels and transporting (16)_____. Scientists have concluded, however, that nutrient (17)_____ do not enhance the performance of well-nourished athletes or active people. Vitamin (18)_____ is a potent fat-soluble antioxidant that defends cell membranes against oxidative damage and physically active people can benefit by using (19)_____ oils and generous servings of fruits and vegetables. Physically active young women, especially those who engage in endurance activities, are prone to (20)_____ deficiency. Early in training, athletes may develop low blood hemoglobin, called (21)_____, but this condition is not a true iron deficiency.

Endurance athletes can lose (22)_____ or more quarts of fluid in every hour of activity and must hydrate before, and rehydrate (23)_____ and after activity to replace it all. Plain, cool (24)_____ is the best fluid for most active bodies, although endurance athletes also need (25)_____ to supplement their limited glycogen stores.

Athletes need a diet composed mostly of (26)_____-dense foods, the kind that supply a maximum of vitamins and minerals for the energy they provide. A diet that is (27)_____ in carbohydrate, moderate in (28)_____, and adequate in (29)_____ ensures full glycogen stores. A diet that provides (30)_____ percent of total calories from carbohydrate, (31)_____ percent from unsaturated fats, and (32)_____ percent from protein works best.

Chapter Glossary

Matching Exercise:

_____ 1. overload

_____ 2. stroke volume

_____ 3. hourly sweat rate

_____ 4. training

_____ 5. muscle endurance

_____ 6. flexibility

_____ 7. cardiorespiratory endurance

_____ 8. weight training

_____ 9. hypertrophy

_____ 10. cardiac output

_____ 11. aerobic

_____ 12. anaerobic

_____ 13. VO$_2$ max

_____ 14. atrophy

a. the ability of a muscle to contract repeatedly within a given time without becoming exhausted

b. an increase in size (for example, of a muscle) in response to use

c. not requiring oxygen

d. the amount of weight lost plus fluid consumed during exercise per hour

e. the ability to perform large muscle dynamic exercise of moderate to high intensity for prolonged periods

f. a decrease in size (for example, of a muscle) because of disuse

g. an extra physical demand placed on the body

h. requiring oxygen

i. the amount of oxygenated blood ejected from the heart toward body tissues at each beat

j. the capacity of the joints to move through a full range of motion; the ability to bend and recover without injury

k. the volume of blood discharged by the heart each minute

l. regular practice of an activity, which leads to physical adaptations of the body, with improvements in flexibility, strength, or endurance

m. the maximum rate of oxygen consumption by an individual (measured at sea level)

n. the use of free weights or weight machines to provide resistance for developing muscle strength and endurance

Crossword Puzzle:

Word Bank:		
exercise	hyponatremia	muscle strength
fitness	hypothermia	physical activity
glucose polymers	lactate	pregame meal
heat stroke	loading	

Across:

1. a meal eaten three to four hours before athletic competition
5. compounds that supply glucose, not as single molecules, but linked in chains somewhat like starch
6. bodily movement produced by muscle contractions that substantially increase energy expenditure
7. an acute and life-threatening reaction to heat buildup in the body
8. planned, structured, and repetitive bodily movement that promotes or maintains physical fitness
9. a regimen of moderate exercise, followed by eating a high-carbohydrate diet, that enables muscles to temporarily store glycogen beyond their normal capacity
10. the characteristics that enable the body to perform physical activity
11. a decreased concentration of sodium in the blood

Down:

2. the ability of muscles to work against resistance
3. a compound produced during the breakdown of glucose in anaerobic metabolism
4. a below-normal body temperature

Exercises

Answer these chapter study questions:

1. Why do some athletes take megadoses of vitamin E?

2. Discuss possible causes of iron deficiency in young female athletes.

3. What is the purpose of carbohydrate loading? Describe the steps involved.

4. What are two routes of significant water loss from the body during exercise?

5. What is the best fluid for most exercising bodies? Why?

6. How would you respond to an athlete who believes that taking vitamin or mineral supplements just before competition will enhance performance?

Complete these short answer questions:

1. The four components of fitness are:

   a.   c.

   b.   d.

2. Benefits of weight training include:

   a.

   b.

   c.

   d.

   e.

   f.

3. These vitamins are important for athletes because they build the red blood cells that carry oxygen to working muscles:

   a.   b.

4. Cardiorespiratory endurance is characterized by:

   a.

   b.

   c.

   d.

   e.

   f.

5. Four strategies which endurance athletes use to try to maintain their blood glucose concentrations as long as they can include:

   a.

   b.

   c.

   d.

6. Factors that affect fat use during physical activity include:

    a.   c.

    b.

7. During physical activity, the body loses these electrolytes in sweat:

    a.   c.

    b.

8. Factors that affect how much protein is used during physical activity include:

    a.   c.

    b.

9. Precautions to prevent heat stroke include:

    a.

    b.

    c.

10. The pregame meal should be eaten _____ hours before athletic competition.

Solve these problems:

1. An athlete weighs 150 pounds prior to a workout and 145 pounds after a workout. How much fluid should be consumed in order to rehydrate the body?

2. Review the example pregame meal below, determine all of the foods and/or beverages which would not be recommended, and identify an appropriate substitute for each non-recommended item.

	Substitute
orange juice	_____
3-ounce hamburger patty	_____
bun	_____
potato chips	_____
raw carrot sticks	_____
frozen yogurt	_____

Answer these controversy questions:

1. Explain why large doses of amino acid supplements can be dangerous.

2. What are some of the possible adverse effects of caffeine consumption among athletes?

3. What is the most appropriate way for an athlete to use a specialty drink or bar?

4. Name groups who have banned the use of androstenedione and DHEA in competitions.

5. Why is carnitine supplementation not necessary for an athlete?

Study Aids

For questions 1-5, match the activities, listed on the left, with the type of fitness they develop, listed on the right.

_____	1. weight lifting		a.	flexibility
_____	2. repetitive exercises like push-ups		b.	muscle strength and endurance
_____	3. stretches		c.	cardiorespiratory endurance
_____	4. swimming			
_____	5. fast bicycling			
_____	6. yoga			

For questions 7-11, match the vitamins and minerals, listed on the right, with their exercise-related functions, listed on the left.

_____	7. helps build the red blood cells that carry oxygen to working muscles		a.	iron
_____	8. helps make muscles contract		b.	folate
_____	9. collagen formation for joint and other tissue integrity		c.	calcium
_____	10. transport of oxygen in blood and in muscle tissue		d.	vitamin C
_____	11. defends cell membranes against oxidation damage		e.	vitamin E

12. Use Table 10-1 on page 371 of your textbook to study the latest guidelines for physical fitness for adults, and study Table 10-5 on page 385 to identify the hydration schedule for physical activity.

Sample Test Items

Comprehension Level Items:

1. The ability to bend and recover without injury is called:

    a. fitness.
    b. flexibility.

    c. strength.
    d. endurance.

2. People who regularly engage in moderate physical activity live longer on average than those who are physically inactive.

    a. true

    b. false

3. The average resting pulse rate for adults is around _____ beats per minute.

    a. 20
    b. 50

    c. 70
    d. 90

4. Which of the following produce cardiorespiratory endurance?

   a. swimming
   b. slow walking
   c. basketball
   d. a and b
   e. a and c

5. Sports anemia:

   a. is the same as iron-deficiency anemia.
   b. reflects a normal adaptation to endurance training.
   c. hinders the ability to perform work.
   d. requires iron supplementation.

6. Athletes relying on thirst to govern fluid intake can easily become dehydrated.

   a. true
   b. false

7. Plain, cool water is the optimal beverage for replacing fluids in those who exercise because it:

   a. rapidly leaves the digestive tract.
   b. takes a long time to enter the tissues.
   c. cools the body from the inside out.
   d. a and b
   e. a and c

8. Which of the following beverages is recommended for an athlete?

   a. iced tea
   b. beer
   c. coffee
   d. fruit juice

9. To postpone fatigue, endurance athletes should:

   a. eat a high-carbohydrate diet on a day-to-day basis.
   b. take in some glucose during the activity, usually in fluid.
   c. avoid carbohydrate-rich foods following exercise.
   d. a and b
   e. b and c

10. Which of the following is an effect of regular physical activity?

    a. decreased lean body tissue
    b. less bone density
    c. reduced HDL cholesterol
    d. reduced risk of cardiovascular disease

11 Cardiorespiratory endurance is characterized by:

    a. increased heart strength and stroke volume.
    b. improved circulation.
    c. increased blood pressure.
    d. a and b
    e. b and c

12. The *Dietary Guidelines for Americans 2005* recommend that people spend an accumulated minimum of 30 minutes in some sort of physical activity on most days of the week.

    a.   true   b.   false

13. Glycogen depletion usually occurs after about _____ hour(s) of vigorous exercise.

    a.   ½   c.   1 ½
    b.   1   d.   2

14. During intense activity, anaerobic breakdown of glucose produces:

    a.   lactate.   c.   amino acids.
    b.   ketone bodies.   d.   glycogen.

15. During events lasting longer than 3 hours, athletes need to pay special attention to replacing sodium losses to prevent hyponatremia.

    a.   true   b.   false

16. After about _____ minutes of activity, the blood fatty acid concentration rises and surpasses the normal resting concentration.

    a.   10   c.   30
    b.   20   d.   45

17. All of the following factors affect fat use in exercise **except**:

    a.   amount of fat in the diet.
    b.   degree of training to perform the exercise.
    c.   intensity and duration of the exercise.
    d.   area of the body involved in the exercise.

18. Which of the following signals the synthesis of muscle proteins?

    a.   dietary protein   c.   physical activity
    b.   adequate nutrients   d.   dietary amino acids

19. All of the following are lost in sweat during physical activity **except**:

    a.   sodium.   c.   potassium.
    b.   calcium.   d.   chloride.

20. Physically fit people have a lower risk of cardiovascular disease due to:

    a.   lower blood pressure.   d.   a and b
    b.   lower HDL cholesterol.   e.   a and c
    c.   lower total blood cholesterol.

21. The Committee on Dietary Reference Intakes recommends at least 60 minutes of moderately intense activity each day to maintain a healthy body weight.

    a.   true   b.   false

22. To emphasize muscle strength in weight training, you would combine less resistance (lighter weight) with more repetitions.

    a. true

    b. false

23. To improve your flexibility you would engage in:

    a. weight lifting.

    b. jumping rope.

    c. yoga.

    d. skating.

24. The Committee on DRI recommends greater-than-normal protein intake for athletes.

    a. true

    b. false

Application Level Items:

25. Based on the DRI intake recommendation for protein for athletes, how many grams of protein should Sally, a long-distance runner, consume if she weighs 130 pounds?

    a. 30

    b. 47

    c. 71

    d. 100

26. A male athlete weighs 175 pounds before a race and 168 pounds after a race. How much fluid should he consume?

    a. 8 cups

    b. 10 cups

    c. 12 cups

    d. 14 cups

27. Approximately how many grams of fat would an endurance athlete need to consume to ensure full glycogen and other nutrient stores and to meet energy needs if she consumes 2500 calories?

    a. 42-58 grams

    b. 56-83 grams

    c. 75-95 grams

    d. 102-130 grams

28. Which of the following foods would be most appropriate as part of a pregame meal?

    a. baked potato with chili and sour cream

    b. marinated vegetable salad with whole-wheat crackers

    c. salami on rye bread with mayonnaise

    d. pasta with steamed vegetables served with French bread

29. Why are people who exercise in humid, hot weather more susceptible to heat stroke?

    a. because they cannot judge how hot they are getting

    b. because they do not sweat under such conditions

    c. because sweat does not evaporate well under such conditions

    d. a and b

    e. b and c

30. Fred wishes to increase his basal metabolic rate. Which of the following should Fred do?

    a. engage in intense, prolonged activity

    b. drink iced tea

    c. eat a high-carbohydrate diet

    d. drink plenty of water

Answers

Summing Up

1. disease
2. strength
3. muscle
4. overload
5. hypertrophy
6. aerobic
7. lean body mass
8. muscle
9. glycogen
10. intensity
11. two
12. duration
13. fat
14. carbohydrate
15. energy
16. oxygen
17. supplements
18. E
19. vegetable
20. iron
21. sports anemia
22. 1.5
23. during
24. water
25. carbohydrate
26. nutrient
27. high
28. unsaturated fat
29. protein
30. 60-70
31. 20-30
32. 10-20

Chapter Glossary

Matching Exercise:
1. g
2. i
3. d
4. l
5. a
6. j
7. e
8. n
9. b
10. k
11. h
12. c
13. m
14. f

Crossword Puzzle:
1. pregame meal
2. muscle strength
3. lactate
4. hypothermia
5. glucose polymers
6. physical activity
7. heat stroke
8. exercise
9. loading
10. fitness
11. hyponatremia

Exercises

Chapter Study Questions:

1. During prolonged, high-intensity physical activity, the muscles' consumption of oxygen increases tenfold or more, enhancing the production of damaging free radicals in the body. Vitamin E is a potent fat-soluble antioxidant that vigorously defends the cell membranes against oxidative damage.

2. Low intakes of iron-rich foods, high iron losses through menstruation, and high demands of muscles for iron-containing molecules of aerobic metabolism and the muscle protein myoglobin can all contribute to iron deficiency.

3. It is a technique used to trick the muscles into storing extra glycogen before a competition. During the week prior to competition, the athlete tapers training and then eats a diet that is high in carbohydrates during the three days before competition.

4. Water is lost from the body via sweat, which increases during exercise in order to rid the body of its excess heat by evaporation. In addition, breathing costs water, exhaled as vapor.

5. Plain, cool water because it rapidly leaves the digestive tract to enter the tissues, where it is needed, and it cools the body from the inside out.

6. Vitamins or minerals taken right before an event do not improve performance. Most vitamins and minerals function as small parts of larger working units and after entering the blood, they have to wait for the cells to combine them with their appropriate other parts so that they can do their work. This may take hours or even days.

Short Answer Questions:

1. (a) flexibility; (b) muscle strength; (c) muscle endurance; (d) cardiorespiratory endurance

2. (a) helps prevent and manage chronic diseases, such as cardiovascular disease; (b) improves older adults' physical mobility; (c) helps maximize and maintain bone mass; (d) enhances psychological well-being; (e) builds lean body mass; (f) helps maintain muscle strength and endurance

3. (a) folate; (b) vitamin B_{12}

4. (a) increased cardiac output and oxygen delivery; (b) increased heart strength and stroke volume; (c) slowed resting pulse; (d) increased breathing efficiency; (e) improved circulation; (f) reduced blood pressure

5. (a) eating a high-carbohydrate diet on a day-to-day basis; (b) taking in some glucose during the activity; (c) training the muscles to store as much glycogen as possible; (d) eating carbohydrate-rich foods following activity to boost storage of glycogen

6. (a) fat intake; (b) intensity and duration of the activity; (c) degree of training

7. (a) sodium; (b) potassium; (c) chloride

8. (a) carbohydrate intake; (b) intensity and duration of the activity; (c) degree of training

9. (a) drinking enough fluid before and during the activity; (b) resting in the shade when tired; (c) wearing light-weight clothing that allows sweat to evaporate

10. 3-4

Problem-Solving:

1. One pound of body weight equals roughly two cups of fluid; since 5 pounds of weight were lost, approximately 10 cups of fluid should be consumed.

2.

	Substitute
orange juice	
3 ounce hamburger patty	omit hamburger patty
bun	dinner roll
potato chips	baked potato
raw carrot sticks	steamed green beans
frozen yogurt	low-fat frozen yogurt

Controversy Questions:

1. Amino acids compete for carriers, and an overdose of one can limit the availability of some other needed amino acids, setting up both a possible toxicity of the supplemented one and a deficiency of the others. Supplements can also lead to digestive disturbances and can cause a buildup of plasma ammonia concentrations.

2. Adverse effects include stomach upset, nervousness, irritability, headaches, dehydration, and diarrhea; it can also raise blood pressure above normal, making the heart work harder to pump blood to the working muscles.

3. The most appropriate way for an athlete to use specialty drinks and bars is as a pregame meal or between-meal snack, especially for a nervous athlete who cannot tolerate solid food on the day of an event.

4. The National Collegiate Athletic Association, the National Football League, and the International Olympic Committee.

5. Because carnitine is a nonessential nutrient that the body makes for itself, plus it is contained in meat and milk products.

Study Aids

1. b	4. c	7. b	10. a
2. b	5. c	8. c	11. e
3. a	6. a	9. d	

Sample Test Items

1. b (p. 370)	9. d (pp. 377-378)	17. d (pp. 379-380)	25. b (p. 382)
2. a (p. 368)	10. d (p. 369)	18. c (p. 381)	26. d (p. 385)
3. c (p. 373)	11. d (p. 373)	19. b (p. 386)	27. b (p. 379)
4. e (p. 373)	12. a (p. 369)	20. e (p. 369)	28. d (p. 391)
5. b (p. 384)	13. d (p. 377)	21. a (pp. 369-370)	29. c (p. 384)
6. a (p. 385)	14. a (p. 376)	22. b (p. 372)	30. a (p. 380)
7. e (p. 386)	15. a (p. 386)	23. c (p. 370)	
8. d (p. 387)	16. b (p. 380)	24. b (p. 382)	

Chapter 11 - Diet and Health

Chapter Objectives

After completing this chapter, you should be able to:

1. Describe the role of nutrition in maintaining a healthy immune system.

2. Define atherosclerosis and describe how it develops.

3. Identify risk factors for cardiovascular disease (CVD), and describe characteristics of a diet to reduce CVD risk.

4. Describe how hypertension develops and identify the risk factors associated with this disease.

5. Discuss how nutrition affects hypertension.

6. Describe the process by which cancer develops.

7. Explain what is known about dietary factors that influence a person's risk of developing cancer.

8. Describe dietary guidelines for disease prevention.

9. Explain the primary causes of the obesity epidemic and how it can be reversed (Controversy 11).

Key Concepts

✓ Adequate nutrition is a key component in maintaining a healthy immune system to defend against infectious diseases. Both deficient and excessive nutrients can harm the immune system.

✓ The same diet and lifestyle risk factors may contribute to several degenerative diseases. A person's family history and laboratory test results can reveal strategies for disease prevention.

✓ Plaques of atherosclerosis trigger abnormal blood clotting and induce hypertension, leading to heart attacks or strokes. Atherosclerosis and hypertension worsen each other.

✓ Major risk factors for CVD put forth by the National Cholesterol Education Program include age, gender, family history, high LDL cholesterol and low HDL cholesterol, hypertension, diabetes, obesity, physical inactivity, smoking, and an atherogenic diet. Other potential risk factors are under investigation.

✓ Diet and exercise are key factors in supporting heart health. High blood cholesterol indicates a risk of heart disease, and diet can contribute to lowering blood cholesterol and thus the risk of heart disease. Dietary measures to lower LDL cholesterol include reducing intakes of saturated fat, *trans* fat, and cholesterol, along with obtaining the fiber, nutrients, and phytochemicals of fruits, vegetables, legumes, and whole grains.

✓ Hypertension is silent, progressively worsens atherosclerosis, and makes heart attacks and strokes likely. All adults should know their blood pressure.

✓ Atherosclerosis, obesity, age, family background, and race contribute to hypertension risks, as do salt intake and other dietary factors, including alcohol.

- For most people, maintaining a healthy body weight, engaging in regular physical activity, minimizing salt and sodium intakes, limiting alcohol intake, and a diet high in fruits, vegetables, fish and low-fat dairy products and low in fat work together to keep blood pressure normal.

- Cancer arises from genetic damage and develops in steps including initiation and promotion, which are thought to be sometimes influenced by diet. The body is equipped to handle tiny doses of carcinogens that occur naturally in foods. Contaminants and naturally occurring toxins can be carcinogenic, but they are monitored in the U.S. food supply.

- Obesity, alcohol consumption, and diets high in certain fats and red meats are associated with cancer development. Foods containing ample fiber, folate, calcium, many other vitamins and minerals, and phytochemicals, along with an ample intake of fluid, are thought to be protective.

Summing Up

The diseases that afflict people around the world are of two main kinds, infectious disease and (1)_____ disease. Nutrition can help strengthen the body's defenses against (2)_____ diseases by supporting a healthy immune system. With respect to prevention of (3)_____ diseases, life choices people make can help to at least postpone them and sometimes to avoid them altogether.

The body's (4)_____ system guards continuously against disease-causing agents. Impaired immunity opens the way for diseases, diseases impair (5)_____ assimilation, and nutrition status suffers further. (6)_____ is especially destructive to various immune system organs and tissues.

People with AIDS and cancer frequently experience malnutrition and (7)_____ away of the body's tissues. Nutrients cannot cure or reverse the progression of AIDS, but an adequate diet may help to improve responses to (8)_____ therapy, reduce duration of (9)_____ stays, and promote greater (10)_____, with an improved quality of life overall.

In contrast to the infectious diseases, the (11)_____ diseases of adulthood tend to have clusters of suspected contributors known as (12)_____. Among them are environmental, (13)_____, social, and genetic factors that tend to occur in clusters and interact with each other. A person who eats a diet too high in saturated fat, salt and calories increases his or her probabilities of becoming (14)_____ and of contracting cancer, hypertension, diabetes, atherosclerosis, diverticulosis or other diseases. Many experts believe that diet accounts for about a (15)_____ of all cases of coronary heart disease and general trends support the link between diet and (16)_____.

To decide whether certain (17)_____ recommendations are especially important to you, you should consider your family's (18)_____ history to see which diseases are common to your forebears. The combination of family history and (19)_____ test results is a powerful predictor of disease.

In addition to diet, age, gender, genetic inheritance, cigarette smoking, certain diseases and (20)_____ predict CVD development. Some of the major diet-related risk factors for CVD are (21)_____, hypertension, and high blood (22)_____ and low blood HDL. High (23)_____ and low HDL correlate directly with risk of heart disease. An atherogenic diet, which is high in saturated fat, *trans* fats, and (24)_____, increases LDL cholesterol.

In addition to atherosclerosis, several major risk factors predict the development of hypertension, including (25)_____, obesity, genetics, alcohol, and (26)_____ intake. A blanket recommendation for prevention of hypertension would center on controlling (27)_____, consuming a nutritious diet, exercising regularly, controlling alcohol intake, and holding sodium intake to prescribed levels. People with kidney problems or diabetes, those with a family history of hypertension, African Americans, and persons who are older are especially likely to respond more sensitively than others to (28)_____ intakes. One diet which consistently improves blood pressure in many people who follow it is called the (29)_____ diet.

An estimated (30)_____ percent of cancers are influenced by diet. Studies suggest diets high in certain fats and (31)_____, are associated with (32)_____ development. Foods containing ample fiber, folate, calcium, many other vitamins and minerals, and (33)_____, along with an ample intake of (34)_____, are thought to be protective.

Chapter Glossary

Matching Exercise:

_____ 1. anticarcinogens

_____ 2. metabolic syndrome

_____ 3. hypertension

_____ 4. carcinogen

_____ 5. plaques

_____ 6. thrombus

_____ 7. metastasis

_____ 8. cruciferous vegetables

_____ 9. atherosclerosis

_____ 10. cancer

_____ 11. calorie effect

_____ 12. heart attack

_____ 13. embolus

_____ 14. aneurysm

_____ 15. initiation

_____ 16. platelets

_____ 17. risk factors

_____ 18. stroke

_____ 19. promoters

_____ 20. acrylamide

_____ 21. macrophages

a. a disease in which cells multiply out of control and disrupt normal functioning of one or more organs

b. high blood pressure

c. tiny cell-like fragments in the blood, important in blood clot formation

d. a stationary blood clot

e. factors that do not initiate cancer but speed up its development once initiation has taken place

f. vegetables with cross-shaped blossoms that are associated with low cancer rates

g. a thrombus that breaks loose and travels through the blood vessels

h. compounds in foods that act in several ways to oppose the formation of cancer

i. the event in which the vessels that feed the heart muscle become closed off by an embolism, thrombus, or other cause with resulting sudden tissue death

j. the sudden shutting off of the blood flow to the brain by a thrombus, embolism, or the bursting of a vessel

k. a cancer-causing substance

l. an event, probably occurring in the cell's genetic material, caused by radiation or by a chemical carcinogen that can give rise to cancer

m. a combination of characteristic factors—central obesity, high fasting blood glucose or insulin resistance, hypertension, low blood HDL cholesterol and elevated triglycerides—that greatly increase a person's risk of developing CVD

n. the drop in cancer incidence seen whenever intake of food energy is restricted

o. factors known to be related to (or correlated with) diseases but not proved to be causal

p. mounds of lipid material, mixed with smooth muscle cells and calcium, that develop in the artery walls in atherosclerosis

q. the ballooning out of an artery wall at a point that is weakened by deterioration

r. the most common form of cardiovascular disease, characterized by plaques along the inner walls of the arteries

s. movement of cancer cells from one body part to another, usually by way of the body fluids

t. large scavenger cells of the immune system that engulf debris and remove it

u. a chemical produced in carbohydrate-rich foods, such as potatoes and grains, cooked at high temperatures

Crossword Puzzle:

Word Bank:

AIDS	degenerative diseases	infectious diseases
aorta	diastolic pressure	systolic pressure
bioterrorism	embolism	thrombosis
carcinogenesis		

	Across:		*Down:*
1.	the second figure in a blood pressure reading, which reflects the arterial pressure when the heart is between beats	1.	chronic, irreversible diseases characterized by degeneration of body organs due in part to such personal lifestyle elements as poor food choices, smoking, alcohol use, and lack of physical activity
4.	an embolus that causes sudden closure of a blood vessel	2.	the origination or beginning of cancer
7.	a thrombus that has grown enough to close off a blood vessel	3.	the first figure in a blood pressure reading, which reflects arterial pressure caused by the contraction of the heart's left ventricle
8.	diseases that are caused by bacteria, viruses, parasites, and other microbes and can be transmitted from one person to another through air, water, or food; by contact; or through vector organisms such as mosquitoes or fleas	5.	the intentional spreading of disease-causing organisms or agricultural pests as a political weapon to produce fear and intimidation
9.	the large, primary artery that conducts blood from the heart to the body's smaller arteries	6.	acquired immune deficiency syndrome; caused by infection with human immunodeficiency virus (HIV), which is transmitted primarily by sexual contact, contact with infected blood, needles shared among drug users, or fluids transferred from an infected mother to her fetus or infant

Exercises

Answer these chapter study questions:

1. Why do malnourished children have repeated digestive tract infections?

2. Why must careful attention be paid to food safety when serving food to people with AIDS?

3. How would you go about deciding which diet recommendations and lifestyle changes are especially important to you?

4. What is the difference between a heart attack and a stroke?

5. How does physical activity improve the odds against heart disease?

6. Describe the primary features of metabolic syndrome. What is it associated with?

Complete these short answer questions:

1. These groups of people are especially likely to be caught in the downward spiral of malnutrition and weakened immunity:

    a.

    b

    c.

    d.

2. The major categories of risk factors for degenerative diseases of adulthood include:

    a.   c.

    b.   d.

3. The major risk factors for CVD that cannot be modified include:

    a.

    b.

    c.

4. The primary risk factors that predict the development of hypertension are:

   a.   e.

   b.   f.

   c.   g.

   d.   h.

5. The steps in cancer development are thought to be:

   a.

   b

   c.

   d.

   e.

   f.

6. Alcohol intake is associated with cancers of the:

   a.   c.

   b.

7. The primary features of metabolic syndrome include:

   a.

   b.

   c.

   d.

   e.

8. Examples of cruciferous vegetables include:

   a.   d.

   b   e.

   c.   f.

9. These types of meats should be consumed only occasionally to minimize risks from carcinogens formed during cooking:

   a.   c.

   b.   d.

10. For most people, these nutrition measures work to keep blood pressure normal:

a.

b

c.

d.

e.

Solve these problems:

1. A serving of a food contains 70 calories and 2 grams of fat. Calculate the percentage of calories from fat in a serving of the food. Would this food be appropriate on a diet in which calories from fat would be 30 percent or less?

2. Jane is a 35 year old who is 65″ tall and weighs 130 pounds. She consumes 54 grams of protein, 150 grams of carbohydrate, and 76 grams of fat per day.

a. How many calories does Jane consume?

b. What is the percent of calories from fat in Jane's diet? Does that amount represent an appropriate diet to control dietary lipids based on the *Dietary Guidelines for Americans*?

Answer these controversy questions:

1. Describe general characteristics of an obesity-promoting environment.

2. How have changes in family structure and working habits influenced people's food intakes over the past five decades?

3. What are some of the techniques used to sell foods to youth?

4. Describe characteristics of a combined approach which represents a workable plan for attacking the obesity epidemic.

5. How can monetary incentives be used to encourage food producers and consumers toward healthier food choices?

Study Aids

1. Complete the following table on adult standards for blood pressure, BMI, blood lipids, and risk of CVD.

Blood Pressure	Systolic and diastolic pressure: 120-139/80-89	= _____ risk
	_____/_____	= healthy
Total Cholesterol	Below _____ mg/dL	= healthy
	200-239 mg/dL	= _____
	_____ mg/dL	= unhealthy
L DL Cholesterol	Below _____ mg/dL	= healthy
	130-159 mg/dL	= _____
	_____ mg/dL	= unhealthy
Obesity	Body mass index ≥ _____	= unhealthy
HDL Cholesterol	HDL < _____ mg/dL	= unhealthy
	≥ 60 mg/dL	= _____
Triglycerides (Fasfng)	< _____ mg/dL	= healthy

2. Use Table 11-9 on page 433 of your textbook to study the dietary guidelines for lowering disease risks.

Sample Test Items

Comprehension Level Items:

1. Nutrition can directly prevent or cure infectious diseases.

   a. true   b. false

2. What are the effects of malnutrition on the body's defense systems?

   a. cells of the immune system are reduced in size
   b. antibody concentration becomes depressed
   c. skin becomes thick with more connective tissue
   d. a and b
   e. b and c

3. A deficiency or a toxicity of even a single nutrient can seriously weaken even a healthy person's immune defense.

   a. true   b. false

4. Degenerative diseases arise from a mixture of:

   a. genetic inheritance.
   b. lifestyle choices.
   c. infection.

   d. a and b
   e. a and c

5. Diet is not the only, and perhaps not even the most important, factor in development of CVD.

   a. true

   b. false

6. Which of the following is **not** part of dietary changes recommended to decrease LDL cholesterol?

   a. decrease dietary saturated fat
   b. decrease dietary cholesterol

   c. decrease soluble fiber
   d. decrease *trans* fat

7. To lower LDL and reduce heart disease risk, people in the United States should:

   a. consume no more than 35% of their calories as fat.
   b. consume no more than 20% of their calories as saturated fat.
   c. consume no more than 300 milligrams of cholesterol per day.
   d. a and b
   e. a and c

8. Which of the following correlates directly with a higher risk of heart disease?

   a. high total cholesterol
   b. high LDL cholesterol

   c. high triglycerides
   d. high HDL cholesterol

9. Up to age 45, a woman's risk of developing heart disease is lower than a man's of the same age.

   a. true

   b. false

10. To help lower blood cholesterol you should:

    a. reduce the amount of saturated fat in the diet.
    b. omit eggs from the diet altogether.
    c. choose foods high in soluble fiber.
    d. a and b
    e. a and c

11. The twin demons that worsen cardiovascular disease are:

    a. atherosclerosis.
    b. hypertension.
    c. strokes.

    d. a and b
    e. b and c

12. For people 55 or older, the lifetime risk of developing hypertension approaches _____ percent.

    a. 60
    b. 70

    c. 80
    d. 90

13. The best kind of exercise to reduce hypertension is the prolonged kind, such as walking or jogging, recommended for increasing HDL and lowering LDL.

    a. true

    b. false

14. All of the following are likely to respond more sensitively than others to sodium intakes **except** for:

    a. those of European or Asian descent.
    b. those with kidney problems.
    c. those with diabetes.
    d. those whose parents have hypertension.

15. For people with hypertension, a reduction of blood pressure usually accompanies a lower salt intake.

    a. true
    b. false

16. For people who are overweight and hypertensive, a loss of _____ pounds may significantly lower blood pressure.

    a. 2
    b. 5
    c. 8
    d. 10

17. Which of the following would you do to consume more phytochemicals in your diet?

    a. eat more fruit
    b. take dietary supplements
    c. choose cruciferous vegetables
    d. a and b
    e. a and c

18. To minimize risks from carcinogens formed during cooking meats you would:

    a. bake the food in the oven.
    b. grill the food.
    c. pan-fry the food.
    d. smoke the food.

19. Cruciferous vegetables:

    a. are associated with low cancer rates.
    b. include vegetables such as cauliflower, cabbage, and brussels sprouts.
    c. contain chemicals that act as promoters of cancer.
    d. a and b
    e. b and c

20. The DRI Committee's Tolerable Upper Intake Level for sodium is _____ milligrams per day.

    a. 1200
    b. 1450
    c. 2050
    d. 2300

21. Soluble fibers in such foods as oats, barley and legumes help to improve blood lipids by:

    a. preventing cholesterol from being digested.
    b. removing cholesterol from the food.
    c. reducing the cholesterol in the food.
    d. binding cholesterol and bile in the intestine and reducing absorption.

22. For middle-aged and older people, moderate alcohol consumption has been reported to raise HDL cholesterol concentrations or reduce the risk of blood clots.

    a. true
    b. false

23. Which of the following sources of dietary fats are harmless or confer benefits on the heart?

    a. avocados
    b. fish oil supplements
    c. nuts
    d. a and b
    e. a and c

24. Features of the metabolic syndrome include all of the following **except**

    a. low HDL cholesterol.
    c. low triglycerides.
    b. high blood pressure.
    d. high fasting blood glucose.

Application Level Items:

25. You are offering help to an AIDS patient. To support his nutrition status you would:

    a. provide three large meals per day.
    b. insist that food be consumed rather than supplements.
    c. discourage the use of liquid meal replacements.
    d. thoroughly cook all foods and practice cleanliness.

26. You are concerned that you may be at risk for developing CVD. Which of the following would indicate that you are at risk?

    a. BMI of 30
    b. your father died of a heart attack at age 53
    c. your HDL is 60 mg/dL
    d. a and b
    e. b and c

27. To increase consumption of fiber you would do all of the following **except**

    a. consume more dried peas and beans.
    c. choose more white breads.
    b. eat more fresh fruit.
    d. eat more whole-grain foods.

28. You are a 35-year-old male who has just had a physical exam, the results of which are shown below. Which of these results indicates that you are at risk for CVD?

    a. blood cholesterol of 180 mg/dL
    c. HDL cholesterol of 35 mg/dL
    b. body mass index of 22
    d. blood pressure of 110/70

29. To decrease your risk of cancer you would:

    a. decrease your intake of fat.
    b. increase your intake of smoked foods.
    c. eat foods high in fiber.
    d. a and b
    e. a and c

30. Henry is interested in reducing his risk for developing hypertension. He should do all of the following **except**

    a. lose weight if he is overweight.
    c. decrease his consumption of potassium.
    b. moderate alcohol consumption.
    d. moderately restrict salt intake.

Answers

Summing Up

1. degenerative
2. infectious
3. chronic
4. immune
5. food
6. protein-energy malnutrition
7. wasting
8. drug
9. hospital
10. independence
11. degenerative
12. risk factors
13. behavioral
14. obese
15. third
16. cancer
17. diet
18. medical
19. laboratory
20. physical inactivity
21. diabetes
22. LDL
23. LDL
24. cholesterol
25. age
26. salt
27. weight
28. sodium
29. DASH
30. 20-30
31. red meats
32. cancer
33. phytochemicals
34. fluid

Chapter Glossary

Matching Exercise:

1. h
2. m
3. b
4. k
5. p
6. d
7. s
8. f
9. r
10. a
11. n
12. i
13. g
14. q
15. l
16. c
17. o
18. j
19. e
20. u
21. t

Crossword Puzzle:

1. diastolic pressure (across); degenerative diseases (down)
2. carcinogenesis
3. systolic pressure
4. embolism
5. bioterrorism
6. AIDS
7. thrombosis
8. infectious diseases
9. aorta

Exercises

Chapter Study Questions:

1. The linings, antibody concentration, and number of immune cells normally present in secretions of the digestive tract are depressed in malnutrition which allows infectious agents that would normally be barred from the body to enter. Once they are inside the body, the defense mounted against them is weak.

2. Because a common bacteria in food, such as *Salmonella*, can cause a deadly infection in people with compromised immunity, such as individuals with AIDS.

3. You should consider your family's medical history to see which diseases are common and also notice which test results are out of line during your physical examination. The combination of family medical history and laboratory test results is a powerful predictor of disease.

4. When an embolus lodges in an artery of the heart and causes sudden death of part of the heart muscle, a heart attack occurs. A stroke occurs when the embolus lodges in an artery of the brain and kills a portion of brain tissue.

5. Regular physical activity strengthens the muscles of the heart and arteries and expands the heart's capacity to pump blood to the tissues with each beat, reducing the number of heartbeats required and the workload of the heart. Physical activity also favors lean over fat tissue for a healthy body composition and stimulates development of new arteries to nourish the heart muscles.

6. It consists of central obesity combined with any two or more of the following: low HDL, high blood triglycerides, hypertension, and elevated fasting blood glucose (insulin resistance) or type 2 diabetes. It is a distinct array of risk factors that often occurs with CVD.

Short Answer Questions:

1. (a) people who restrict their food intake; (b) the very young or old; (c) the hospitalized; (d) the poor

2. (a) environmental; (b) behavioral; (c) social; (d) genetic

3. (a) increasing age; (b) male gender; (c) family history of premature heart disease

4. (a) obesity; (b) atherosclerosis; (c) insulin resistance; (d) age; (e) genetics; (f) salt intake; (g) alcohol; (h) other dietary factors

5. (a) exposure to a carcinogen; (b) entry of the carcinogen into a cell; (c) initiation; (d) acceleration of other carcinogens called promoters; (e) spreading of cancer cells via blood and lymph; (f) disruption of normal body functions

6. (a) mouth; (b) throat; (c) breast

7. (a) central obesity; (b) low blood HDL; (c) hypertension; (d) elevated fasting glucose (insulin resistance) or type 2 diabetes; (e) elevated triglycerides

8. (a) cabbage; (b) cauliflower; (c) broccoli; (d) brussels sprouts; (e) greens; (f) rutabagas

9. (a) smoked; (b) grilled; (c) fried; (d) broiled

10. (a) healthy body weight; (b) regular physical activity; (c) moderation in those who use alcohol; (d) diet high in fruits, vegetables, fish and low-fat dairy products, and reduced in fat; (e) reduction in salt and sodium intake

Problem-Solving:

1. 26% calories from fat; yes

2. (a) 1500 calories; (b) percentage calories from fat is 46%; no, because fat should total no more than 35% of calories

Controversy Questions:

1. It is an environment which promotes a sedentary lifestyle and over-consumption of high-energy foods. For example, the over-consumption of high-energy foods results from many factors: (1) people are eating out more often and most restaurants appeal to taste and economy, not nutrition, so restaurant foods may be high in fat and/or sugar; also most restaurants serve large portions of foods and encourage overeating; (2) most people work and have less time for meal preparation and rely more on partially-prepared convenience foods to save time, but often these foods are high in sugar and fat; and (3) food manufacturers promote larger portions of foods to encourage consumers to make large food purchases, which result in larger profits for the company. People have also become more sedentary due to a variety of factors: (1) technology has relieved us of many physical demands and many people who were once employed in physically demanding jobs now sit behind computer

screens all day; (2) we live in automobile dependent communities and have urban sprawl, which encourage driving versus biking or walking; (3) entertainment has moved from active forms of physical activity outside, such as playing sports, to inside sedentary activities such as playing computer games, watching DVDs, or watching television; and (3) as people gain weight and become even more sedentary, they may experience higher stress levels and sleep less, resulting in increased exhaustion, which may lead to increased consumption of high-energy drinks or bars to fight fatigue.

2. Most people work, which means that they have less time to devote to the traditional activities of homemaking, including less meal planning, grocery shopping, and labor-intensive preparation required for meals of fresh foods. As a result, more prepared or partially-prepared foods are used.

3. The techniques include manipulating children's emotional and physical needs, such as needs for peer acceptance, love, safety, security, desire for independence, to act older, or to develop an identity.

4. A combined approach would include policy changes, new research from the scientific community, and actions taken by industries and individuals.

5. Taxes could be raised on high-calorie, high-fat, high-salt foods, sugary sodas, candies, and fried snack foods, and the revenues used to subsidize purchase of healthier food choices. Soon, consumers would respond to the lower prices of healthier foods by buying more of them.

Study Aids

Blood Pressure	Systolic and diastolic pressure: 120-139/80-89	= borderline risk
	≤ 120/ ≤ 80	= healthy
Total Cholesterol	Below 200 mg/dL	= healthy
	200-239 mg/dL	= borderline
	≥240 mg/dL	= unhealthy
LDL Cholesterol	Below 100 mg/dL	= healthy
	130-159 mg/dL	= borderline
	160-189 mg/dL	= unhealthy
Obesity	Body mass index ≥ 30	= unhealthy
HDL Cholesterol	HDL < 40 mg/dL	= unhealthy
	≥ 60 mg/dL	= healthy
Triglycerides (Fasting)	< 150 mg/dL	= healthy

Sample Test Items

1. b (p. 402)
2. d (pp. 403-404)
3. a (p. 404)
4. d (p. 403)
5. a (p. 409)
6. c (pp. 415-416)
7. e (p. 415)
8. b (p. 410)

9. a (p. 409)
10. e (pp. 414-416)
11. d (p. 411)
12. d (p. 419)
13. a (p. 420)
14. a (p. 420)
15. a (p. 420)
16. d (p. 420)

17. e (pp. 430-431)
18. a (p. 429)
19. d (p. 430)
20. d (p. 420)
21. d (p. 416)
22. a (p. 416)
23. e (p. 415)
24. c (p. 414)

25. d (p. 404)
26. d (pp. 409-411)
27. c (p. 416)
28. c (p. 411)
29. e (pp. 428-429)
30. c (pp. 420-421)

Chapter 12 - Food Safety and Food Techology

Chapter Objectives

After completing this chapter, you should be able to:

1. List and prioritize the areas of concern regarding the U.S. food supply identified by the Food and Drug Administration.

2. Describe the various sources and symptoms of foodborne illnesses.

3. Discuss how microbial food poisoning can be prevented and indicate which foods are particularly troublesome.

4. Explain new advances in technology related to microbial food safety.

5. Describe the risks, if any, of the following in foods: natural toxins, pesticides, hormone residues, and environmental contaminants.

6. Discuss the regulations concerning food additives and identify the major functions of the various classes of additives.

7. Identify the different food processing techniques and explain the effect they have on the nutrient content of foods.

8. List the arguments for and against the use of products of biotechnology (Controversy 12).

Key Concepts

✓ Each year in the United States, many millions of people suffer mild to life-threatening symptoms caused by foodborne illness, and some die from these illnesses.

✓ Industry employs sound practices to safeguard the commercial food supply from microbial threats. Still, outbreaks of commercial foodborne illnesses have caused widespread harm to health. Consumers should carefully inspect foods before purchasing them.

✓ Foodborne illnesses are common but most cases can be prevented. To prevent them, remember four "keepers": keep hands and surfaces clean, keep raw foods separate, keep hot foods hot, and keep cold foods cold.

✓ Some special food-safety concerns arise when traveling. To avoid foodborne illnesses, remember to boil it, cook it, peel it, or forget it.

✓ Natural foods contain natural toxins that can be hazardous if consumed in excess. To avoid poisoning by toxins, eat all foods in moderation, treat chemicals from all sources with respect, and choose a variety of foods.

✓ Pesticides can be part of a safe food production process but can also be hazardous if mishandled. The FDA tests for pesticide residues in both domestic and imported foods. Consumers can take steps to minimize their ingestion of pesticide residues in foods.

✓ Bovine somatotropin causes cattle to produce more meat and milk on less feed than untreated cattle and the FDA has deemed products from treated cattle to be safe. Antibiotic overuse fosters antibiotic resistance in bacteria, threatening human health. Arsenic drugs are used to promote growth in chickens and other livestock.

✓ Persistent environmental contaminants pose a significant, but generally small, threat to the safety of food. An accidental spill can create an extreme hazard. Mercury and other contaminants are of greatest concern during pregnancy, lactation, and childhood.

✓ The FDA regulates the use of intentional additives. Additives must be safe, effective, and measurable in the final product. Additives on the GRAS list are assumed to be safe because they have long been used. Approved additives have wide margins of safety.

✓ Microbial food spoilage can be prevented by antimicrobial additives. Of these, sugar and salt have the longest history of use. Nitrites and sulfites have advantages and drawbacks. Among flavor additives, the flavor enhancer MSG causes reactions in people with sensitivities to it.

✓ Incidental additives are substances that get into food during processing. They are well regulated, and most do not constitute a hazard.

Summing Up

The (1)_____ is the major agency charged with monitoring the U.S. food supply. According to the FDA, the food hazard which has the potential to cause the greatest harm to people is (2)_____ foodborne illness.

The term foodborne illness refers either to foodborne infection or to food (3)_____. The symptoms of one neurotoxin stand out as severe and commonly fatal—those of (4)_____. The botulinum toxin is destroyed by (5)_____ so canned foods that have been boiled for ten minutes are generally safe from this threat.

Food can provide ideal conditions for (6)_____ to thrive or to produce toxins. Disease-causing bacteria require warmth, (7)_____, and nutrients. To defeat them, people who prepare food should keep hot food hot, keep cold food cold, keep raw foods separate, and keep their hands and (8)_____ clean. Foods that are high in moisture and nutrients and those that are chopped or (9)_____ are especially favorable hosts for microbial growth. Temperature is the most critical factor in keeping (10)_____ safe to eat and ground meat and (11)_____ should be cooked to the well-done stage. Although meats, eggs and seafood posed the greatest foodborne illness threat ten years ago, today (12)_____ poses a similar degree of threat.

A food (13)_____ is anything that does not belong there, and the potential harmfulness depends in part on the extent to which it lingers in the environment or in the human body. In general, the threat from environmental contaminants is small because the (14)_____ monitors the presence of contaminants in foods. Risks to health from pesticide exposure are probably small for healthy adults, but (15)_____ and infants may be at risk for pesticide poisoning. To avoid natural toxins in foods, consumers should eat all foods in (16)_____ and choose a (17)_____ of foods. To remove pesticide residues on fresh fruits and vegetables, consumers should wash them or choose (18)_____ produce.

Manufacturers use food (19)_____ to give foods desirable characteristics, including color, flavor, (20)_____, stability, enhanced nutrient composition, or resistance to spoilage. The use of additives is regulated by the FDA and additives must be safe, (21)_____ and detected and measured in the final food product. Additives included on the (22)_____ list are assumed to be safe because they have been used for a long time and their use entailed no known hazards.

Preservatives known as (23)_____ agents are used to protect foods from growth of microbes. The two best known and most widely used agents are sugar and (24)_____; both work by withdrawing (25)_____ from the food. Another group of antimicrobial agents, the (26)_____, are added to meats and meat products to preserve their color, to inhibit rancidity, and to protect against bacterial growth, especially the deadly (27)_____ bacterium.

Some foods go bad by undergoing changes in color and flavor caused by exposure to (28)_____. (29)_____ are used to prevent oxidation in many processed foods, in alcoholic beverages, and in drugs. However, the FDA prohibits sulfite use on food meant to be eaten (30)_____. A well-known flavor enhancer, (31)_____, is an example of an additive widely used in restaurants, especially Asian restaurants.

In general, the more heavily foods are processed, the less (32)_____ they become. For example, processed foods often gain sodium, as needed (33)_____ is leached away. However, in modern commercial processing, losses of vitamins seldom exceed (34)_____ percent. In contrast, losses in the (35)_____ percent range occur during food preparation at home.

Chapter Glossary

Matching Exercise:

_____ 1. incidental additives

_____ 2. persistent

_____ 3. sushi

_____ 4. irradiation

_____ 5. contaminant

_____ 6. MSG symptom complex

_____ 7. residues

_____ 8. World Health Organization

_____ 9. GRAS list

_____ 10. outbreak

_____ 11. pesticides

_____ 12. additives

_____ 13. margin of safety

_____ 14. toxicity

_____ 15. ultrahigh temperature

_____ 16. organic foods

_____ 17. cross-contamination

a. the acute, temporary, and self-limiting reactions experienced by sensitive people upon ingesting a large dose of MSG

b. two or more cases of foodborne illness arising from an identical organism acquired from a common source within a limited timeframe

c. substances that are added to foods, but are not normally consumed by themselves as foods

d. the ability of a substance to harm living organisms

e. a Japanese dish that consists of vinegar-flavored rice, seafood, and colorful vegetables, typically wrapped in seaweed

f. an agency of the United Nations charged with improving human health and preventing or controlling diseases in the world's people

g. in reference to food additives, a zone between the concentration normally used and that at which a hazard exists

h. the application of ionizing radiation to foods to reduce insect infestation or microbial contamination, or to slow the ripening or sprouting process

i. substances that can get into food not through intentional introduction but as a result of contact with food during growing, processing, packaging, storing, or some other stage before the food is consumed

j. a list of food additives, established by the FDA, long in use and believed safe

k. a process of sterilizing food by exposing it for a short time to temperatures above those normally used in processing

l. whatever remains

m. chemicals used to control insects, diseases, weeds, fungi, and other pests on crops and around animals

n. of a stubborn or enduring nature

o. any substance occurring in food by accident; not a normal food constituent

p. the contamination of a food through exposure to utensils, hands, or other surfaces that were previously in contact with a contaminated food

q. foods meeting USDA production regulations which include prohibition of synthetic pesticides, herbicides, fertilizers, drugs, and preservatives and produced without genetic engineering or irradiation

Crossword Puzzle:

Word Bank:

bacteriophage
bioaccumulation
biotechnology
botulism
bovine somatotropin
enterotoxins

foodborne illness
growth hormone
hazard
Hazard Analysis Critical Control Point
heavy metal

microbes
neurotoxins
pasteurization
prion
tolerance limit

Across:	*Down:*
2. a virus that infects and destroys bacteria	1. an infective agent consisting of an unusually folded protein that disrupts normal cell functioning, causing disease
4. a hormone (somatotropin) that promotes growth and that is produced naturally in the pituitary gland of the brain	3. poisons that act upon mucous membranes such as those of the digestive tract
7. the treatment of milk with heat sufficient to kill certain pathogens commonly transmitted through milk; not a sterilization process	5. a systematic plan to identify and correct potential microbial hazards in the manufacturing, distribution, and commercial use of food products (first 2 of 5 words in the answer)
9. any of a number of mineral ions such as mercury and lead	
12. the maximum amount of a residue permitted in a food when a pesticide is used according to label directions	6. a systematic plan to identify and correct potential microbial hazards in the manufacturing, distribution, and commercial use of food products (last 3 of 5 words in the answer)
13. poisons that act upon the cells of the nervous system	
14. an often-fatal food poisoning caused by botulinum toxin, a toxin produced by bacteria that grow without oxygen in nonacidic canned foods	8. the accumulation of a contaminant in the tissues of living things at higher and higher concentrations along the food chain
15. growth hormone of cattle which can be produced for agricultural use by genetic engineering	9. a state of danger; used to refer to any circumstance in which harm is possible under normal conditions of use
16. illness transmitted to human beings through food and water	10. minute organisms too small to observe without a microscope, including bacteria, viruses, and others
	11. the science that manipulates biological systems or organisms to modify their products or components or create new products

Exercises

Answer these chapter study questions:

1. How can consumers practice food safety and help protect themselves in the grocery store when purchasing canned and packaged foods?

2. Identify the steps which the FDA has taken to protect people who are allergic to sulfites.

3. Why do some foods encourage microbial growth more than others?

4. Explain what is meant by the statement that food processing involves a trade-off.

5. Why does raw produce pose a major foodborne illness threat today?

6. What is the Hazard Analysis Critical Control Point (HACCP) plan?

Complete these short answer questions:

1. Manufacturers use food additives to give foods desirable characteristics related to:

   a.   d.

   b.   e.

   c.   f.

2. The best known and most widely used antimicrobial agents are:

   a.   b.

3. Nitrites are added to foods to:

 a.

 b.

 c.

4. Pasteurized egg products should not be consumed by:

 a.

 b.

 c.

 d.

5. Incidental food additives are substances that find their way into food as the result of some phase of:

 a.   d.

 b.   e.

 c.

6. Warning signs of botulism include:

 a.   c.

 b.   d.

7. The 2, 2, and 4 rules of leftover food safety include:

 a.

 b.

 c.

8. To avoid foodborne illnesses when traveling remember to:

 a.   c.

 b.   d.

9. These raw foods are especially likely to contain illness-causing bacteria:

 a.   c.

 b.

10. Four things which should be done in the kitchen in order to prevent food poisoning include:

 a.

 b.

 c.

 d.

Solve these problems:

For questions 1-4, match the types of food additives on the left with examples of the additives on the right.

_____	1. antimicrobial agent	a.	MSG
_____	2. antioxidant	b.	sulfites
_____	3. artificial flavor	c.	iodine
_____	4. nutrient additive	d.	nitrites

For questions 5-8, for each food processing method listed on the left, identify the effects on nutrient content, listed on the right.

_____	5. canning	a.	leaves most nutrients intact
_____	6. freezing	b.	causes substantial losses of water-soluble vitamins
_____	7. drying	c.	considerable nutrient losses, notably vitamin E, fiber, magnesium, and water-soluble vitamins
_____	8. extruding	d.	negligible effects on nutrients

Answer these controversy questions:

1. What is the position of the American Dietetic Association related to agricultural and food biotechnology?

2. What is the primary concern related to the nutrient composition of GE foods?

3. Describe how genetic engineering can result in greater crop yields.

4. What is the FDA's position on cloned animals?

5. Describe the environmental effects of GE crops.

Study Aids

Study Table 12-6 on page 455 of your textbook to determine the safe refrigeration storage times for various foods, and determine ways to reduce pesticide residue intake by studying Table 12-8 on page 465 of your textbook.

Sample Test Items

Comprehension Level Items:

1. Which of the following appears as the number one area of concern on the Food and Drug Administration's list of concerns regarding our food supply?

   a. pesticide residues
   b. microbial foodborne illness
   c. intentional food additives
   d. naturally occurring toxicants

2. Substances that find their way into food as a result of production, processing, storage, or packaging are called _____ additives.

   a. incidental
   b. contaminant
   c. intentional
   d. direct

3. The FDA's authority over additives hinges primarily on their:

   a. appropriate uses.
   b. effectiveness.
   c. safety.
   d. a and b
   e. b and c

4. If the FDA approves an additive's use, that means the manufacturer can:

   a. use it only in amounts specified by the FDA.
   b. add it to only those foods delineated by the FDA.
   c. add it in any amount to any food.
   d. a and b
   e. b and c

5. Which of the following is(are) true concerning the GRAS list?

    a. Some 700 substances have been placed on the list.
    b. Substances on the GRAS list can be re-evaluated.
    c. Substances on the GRAS list cannot be changed.
    d. a and b
    e. b and c

6. MSG has been deemed safe for adults, but is kept out of foods for infants.

    a. true   b. false

7. Salt and sugar work as antimicrobial agents by:

    a. removing oxygen from foods.
    b. withdrawing nutrients used as food by bacteria.
    c. withdrawing water from food.
    d. b and c
    e. a and b

8. Preservatives known as _____ prevent or delay rancidity of fats.

    a. glutamates   c. antimicrobials
    b. tartrazines   d. antioxidants

9. To reduce pesticide residues in foods you would:

    a. trim the fat from meat.
    b. consume waxed vegetables without removing the wax.
    c. wash fresh produce in warm water, using a scrub brush.
    d. a and b
    e. a and c

10. No food additives are permanently approved by the FDA.

    a. true   b. false

11. To prevent bacterial growth, you should refrigerate foods immediately after serving a meal and definitely before _____ has(have) passed.

    a. 30 minutes   c. 2 hours
    b. 1 hour   d. 3 hours

12. Which of the following would be the safest food to pack for a picnic?

    a. apples   c. chicken salad
    b. pasta salad   d. Swiss cheese

13. Which of the following processes causes substantial losses of water-soluble vitamins?

    a. canning   c. drying
    b. freezing   d. ultrahigh temperature

14. Which of the following is a special enemy of thiamin?

    a. MSG
    b. pesticides

    c. sulfites
    d. nitrites

15. You can use your senses of smell and sight alone to warn you of hazards and food contamination.

    a. true

    b. false

16. During processing, foods often gain:

    a. potassium.
    b. iron.

    c. calcium.
    d. sodium.

17. When dining in an Asian restaurant, individuals who are sensitive to MSG should:

    a. order a soup that contains noodles.
    b. order plain broth.
    c. eat plenty of rice.

    d. a and b
    e. a and c

18. Wooden cutting boards do not support microbial growth and therefore require no treatment.

    a. true

    b. false

19. To prevent traveler's diarrhea you would:

    a. wash hands before eating.
    b. drink only bottled beverages.
    c. wash fresh fruits under running water.
    d. a and b
    e. b and c

20. Which of the following appears last on the FDA's list of concerns regarding the U.S. food supply?

    a. residues in foods
    b. genetic modification of foods

    c. natural toxins in foods
    d. intentional food additives

21. Foodborne illnesses can be life-threatening and are increasingly unresponsive to standard antibiotics.

    a. true

    b. false

22. Which of the following is(are) characteristic of a safe hamburger?

    a. one cooked until it turns brown
    b. one that has reached an internal temperature of 160° F
    c. one cooked until well-done
    d. a and b
    e. b and c

23. To decrease foodborne illness, you would rinse and scrub all raw produce thoroughly under running water and discard the outer leaves from heads of leafy vegetables, such as cabbage and lettuce, before washing.

    a. true

    b. false

24. The FDA and EPA advise all pregnant women against eating king mackerel, swordfish and shark because they are high in:

    a. lead.   c. selenium.

    b. mercury.   d. arsenic.

Application Level Items:

25. Which of the following food products would be acceptable to purchase from the grocery store?

    a. a jar of jelly with a broken seal   c. a can of tomatoes that is bulging

    b. a solidly frozen package of spinach   d. a torn package of frozen green peas

26. Which of the following foods would be most susceptible to bacterial contamination?

    a. pork chop   c. meatloaf

    b. chicken breast   d. veal cutlet

27. Which of the following foods would be most appropriate to carry on a picnic?

    a. pimento cheese spread   c. egg salad

    b. meatballs   d. canned cheese spread

28. To prevent food poisoning you would:

    a. transfer leftovers to deep containers before refrigerating.

    b. refrigerate perishables immediately when you get home.

    c. thaw poultry in the refrigerator.

    d. a and b

    e. b and c

29. The FDA has approved the use of aspartame as a sweetener. This means that:

    a. it can be added to only certain approved food products.

    b. it can be added in unlimited quantities to foods.

    c. it can be used forever in food products.

    d. it can be placed on the GRAS list.

30. You are interested in using a preservative which will protect a food from microbes and prevent the food from becoming hazardous to health. Which of the following would you use?

    a. ascorbate   c. tartrazine

    b. sulfites   d. nitrites

Answers

Summing Up

1. Food and Drug Administration
2. microbial
3. poisoning
4. botulism
5. heat
6. bacteria
7. moisture
8. surfaces
9. ground
10. meat
11. poultry
12. produce
13. contaminant
14. FDA
15. children
16. moderation
17. variety
18. organic
19. additives
20. texture
21. effective
22. GRAS
23. antimicrobial
24. salt
25. water
26. nitrites
27. botulinum
28. oxygen
29. Sulfites
30. raw
31. MSG
32. nutritious
33. potassium
34. 25
35. 60-75

Chapter Glossary

Matching Exercise:
1. i
2. n
3. e
4. h
5. o
6. a
7. l
8. f
9. j
10. b
11. m
12. c
13. g
14. d
15. k
16. q
17. p

Crossword Puzzle:
1. prion
2. bacteriophage
3. enterotoxins
4. growth hormone
5. Hazard Analysis
6. Critical Control Point
7. pasteurization
8. bioaccumulation
9. heavy metal (across); hazard (down)
10. microbes
11. biotechnology
12. tolerance limit
13. neurotoxins
14. botulism
15. bovine somatotropin
16. foodborne illness

Exercises

Chapter Study Questions:

1. Consumers can check the freshness dates on food packages and avoid those with expired dates; also, they can inspect the seals and wrappers on packages and reject open, torn, leaking, or bulging items.

2. It prohibits sulfite use on food intended to be consumed raw, with the exception of grapes; it requires that sulfite-containing foods and drugs list the additives on their labels to warn that sulfites are present.

3. Because the ideal conditions for bacteria to thrive include warmth, moisture, and nutrients; foods that are high in moisture and nutrients and those that are chopped or ground are especially favorable hosts for microbial growth.

4. Food processing makes food safer, gives it a longer usable lifetime than fresh food, or cuts preparation time; however, the cost is the loss of some vitamins and minerals, which we need to consume more of, and the addition of substances, such as sodium, sugar and fat, which we should consume less of.

5. Because these foods grow close to the ground, which makes bacterial contamination from the soil, animal waste runoff, and organic fertilizers likely. Produce may also be imported from other countries that do not adhere to safe growing or harvesting practices and where contagious diseases are widespread.

6. HACCP is a method of preventing foodborne illnesses at their source. Each slaughterhouse, producer, packer, distributor, and transporter of food identifies critical control points in its procedures where the risk of food contamination is high, and then devises and implements ways to eliminate or minimize it.

Short Answer Questions:

1. (a) color; (b) flavor; (c) texture; (d) stability; (e) resistance to spoilage; (f) enhanced nutrient composition

2. (a) sugar; (b) salt

3. (a) preserve their color; (b) inhibit rancidity; (c) protect against bacterial growth

4. (a) pregnant women; (b) elderly; (c) young children; (d) those suffering from immune dysfunction

5. (a) production; (b) processing; (c) storage; (d) packaging; (e) consumer preparation

6. (a) double vision; (b) weak muscles; (c) difficulty swallowing; (d) difficulty breathing

7. (a) within 2 hours of cooking, refrigerate the food; (b) refrigerate the food in shallow containers about 2 inches deep; (c) use it within 4 days or toss it out

8. (a) boil it; (b) cook it; (c) peel it; (d) or forget it

9. (a) meats; (b) eggs; (c) seafood

10. (a) keep hot foods hot; (b) keep cold foods cold; (c) keep your hands and kitchen surfaces clean; (d) keep raw food separate

Problem-Solving:

1. d   3. a   5. b   7. a
2. b   4. c   6. d   8. c

Controversy Questions:

1. Their position is that biotechnology can enhance the quality, safety, nutritional value, and variety of the food supply, while helping to solve problems of production, processing, distribution and environmental and waste management.

2. That overdoses of nutrients or phytochemicals may pose more danger than reduced content.

3. Both herbicide-resistant and insect-resistant crops are being designed to increase usable food yield per acre of farmed land. Herbicide-resistant crops allow farmers to spray whole fields, where only the weeds die and the food crop remains healthy and strong. Insect-resistant crops make pesticides in the plant tissues themselves, which greatly increases yields.

4. The FDA has deemed the milk and meat from cloned animals and their offspring as safe.

5. Positive effects include reduced pesticide use and soil erosion. Current concerns include damaging wildlife and the accidental cross-pollination of plant pesticide crops with related wild weeds, which would give them a survival advantage.

Sample Test Items

1. b (pp. 444-445)
2. a (p. 475)
3. e (pp. 473-474)
4. d (pp. 473-474)
5. d (p. 473)
6. a (p. 475)
7. c (p. 474)
8. d (p. 472)

9. e (p. 465)
10. a (p. 473)
11. c (p. 453)
12. a (p. 458)
13. a (p. 477)
14. c (p. 474)
15. b (p. 450)
16. d (p. 476)

17. e (p. 474)
18. b (p. 452)
19. d (p. 459)
20. b (pp. 444-445)
21. a (p. 445)
22. e (pp. 454-455)
23. a (p. 457)
24. b (p. 471)

25. b (pp. 449-450)
26. c (pp. 454-455)
27. d (pp. 458-459)
28. e (p. 453)
29. a (p. 473)
30. d (pp. 472-474)

Chapter 13 - Life Cycle Nutrition
Mother and Infant

Chapter Objectives

After completing this chapter, you should be able to:

1. Describe how maternal nutrition before and during pregnancy affects both the development of the fetus and growth of the infant after birth.

2. Discuss maternal physiological adjustments that occur during pregnancy and explain how they influence energy and other nutrient requirements.

3. List the components of weight gain during pregnancy and state how much weight a woman should gain during pregnancy.

4. Explain why abstinence from smoking and drugs, avoiding dieting, and moderation in the use of caffeine are recommended during pregnancy.

5. Explain the effects of alcohol on the development of the fetus and describe the condition known as fetal alcohol syndrome.

6. List the benefits of breastfeeding an infant and indicate the changes a lactating woman needs to make in her diet to promote breastfeeding success.

7. Describe the circumstances under which breastfeeding is not recommended and explain what alternatives exist for proper infant nutrition.

8. Identify nutrient needs of an infant and explain why breast milk is the ideal food for an infant.

9. Plan a timetable and discuss guidelines for feeding foods to an infant from birth to 12 months of age.

10. Describe what is currently known about relationships among childhood obesity and early chronic diseases (Controversy 13).

Key Concepts

✓ Adequate nutrition before pregnancy establishes physical readiness and nutrient stores to support fetal growth. Both underweight and overweight women should strive for appropriate body weights before pregnancy. Newborns who weigh less than 5 ½ pounds face greater health risks than normal-weight babies. The healthy development of the placenta depends on adequate nutrition before pregnancy.

✓ Implantation, fetal development, and early critical periods depend on maternal nutrition before and during pregnancy.

✓ Pregnancy brings physiological adjustments that demand increased intakes of energy and nutrients. A balanced diet that includes more nutrient-dense foods from the five food groups can help to meet these needs.

✓ Due to their key roles in cell reproduction, folate and vitamin B_{12} are needed in large amounts during pregnancy. Folate plays an important role in preventing neural tube defects.

- ✓ All pregnant women, but especially those who are less than 25 years of age, need to pay special attention to ensure adequate calcium intakes. A daily iron supplement is recommended for all pregnant women during the second and third trimesters.

- ✓ Women most likely to benefit from multivitamin-mineral supplements during pregnancy include those who do not eat adequately, those carrying twins or triplets, and those who smoke cigarettes or are alcohol or drug abusers.

- ✓ Food assistance programs such as WIC can provide nutritious food for pregnant women of limited financial means.

- ✓ Weight gain is essential for a healthy pregnancy. A woman's prepregnancy BMI, her own nutrient needs, and the number of fetuses she is carrying help to determine appropriate weight gain.

- ✓ Physically fit women can continue to be physically active throughout pregnancy. Pregnant women should be cautious in their choice of activities.

- ✓ Of all the population groups, pregnant teenage girls have the highest nutrient needs and an increased likelihood of having problem pregnancies.

- ✓ Food cravings usually do not reflect physiological needs, and some may interfere with nutrition. Nausea arises from normal hormonal changes of pregnancy.

- ✓ Abstaining from smoking and other drugs, limiting intake of foods known to contain unsafe levels of contaminants such as mercury, taking precautions against foodborne illness, avoiding large doses of nutrients, refraining from dieting, using artificial sweeteners in moderation, and limiting caffeine use are recommended during pregnancy.

- ✓ Alcohol limits oxygen delivery to the fetus, slows cell division, and reduces the number of cells organs produce. Alcoholic beverages must bear warnings to pregnant women.

- ✓ The birth defects of fetal alcohol syndrome arise from severe damage to the fetus caused by alcohol. Lesser conditions, ARND and ARBD, may be harder to diagnose but also rob the child of a normal life.

- ✓ Abstinence from alcohol is critical to prevent irreversible damage to the fetus.

- ✓ Gestational diabetes and preeclampsia are common medical problems associated with pregnancy. These should be managed to minimize associated risks.

- ✓ The lactating woman needs extra fluid and enough energy and nutrients to make sufficient milk each day. Malnutrition most often diminishes the quantity of the milk produced without altering quality. Lactation may facilitate loss of the extra fat gained during pregnancy.

- ✓ Breastfeeding is not advised if the mother's milk is contaminated with alcohol, drugs or environmental pollutants. Most ordinary infections such as colds have no effect on breastfeeding. Where safe alternatives are available, HIV-infected women should not breastfeed their infants.

- ✓ Infants' rapid growth and development depend on adequate nutrient supplies, including water from breast milk or formula.

- ✓ Breast milk is the ideal food for infants because it provides the needed nutrients in the right proportions and protective factors as well.

- ✓ Infant formulas are designed to resemble breast milk and must meet an AAP standard for nutrient composition. Special formulas are available for premature infants, allergic infants, and others. Formula should be replaced with milk only after the baby's first birthday.

- ✓ Solid food additions to an infant's diet should begin at about six months and should be governed by the infant's nutrient needs and readiness to eat. By one year, the baby should be receiving foods from all food groups.
- ✓ The early feeding of the infant lays the foundation for lifelong eating habits. It is desirable to foster preferences that will support normal development and health throughout life.

Summing Up

Before she becomes pregnant, a woman must establish eating habits that will optimally

(1)_____ both the growing fetus and herself and strive for an appropriate body

(2)_____. An underweight woman who fails to gain adequately during pregnancy is most

likely to bear a baby with a dangerously low (3)_____. Infant birthweight is the most

potent single indicator of an infant's future (4)_____ status. Prepregnancy nutrition

determines whether the woman will be able to grow a healthy (5)_____, which is the only

source of (6)_____ available to the fetus.

During the two weeks following fertilization, minimal growth takes place but it is a

(7)_____ period developmentally. At eight weeks, the fetus has a complete central

nervous system, a beating (8)_____, and a fully formed (9)_____ system.

The gestation period, which lasts approximately (10)_____ weeks, ends with the

(11)_____ of the infant.

During the second trimester, the pregnant woman needs (12)_____ extra calories per day above

the allowance for nonpregnant women. In addition, about (13)_____ more grams of protein are

required. Pregnant women need ample amounts of (14)_____-rich foods to spare their

protein and fuel the fetal brain. Vitamins needed in large amounts during pregnancy, because of their

roles in cell reproduction, include (15)_____ and vitamin B_{12}. Calcium,

(16)_____, and magnesium are the minerals in greatest demand during pregnancy.

Although the body conserves (17)_____ even more than usual during pregnancy, an iron

supplement is recommended during the second and third trimesters. The ideal pattern of weight gain is

thought to be 3 ½ pounds during the first (18)_____ months and a

(19)_____ per week thereafter.

The energy cost of producing milk is about (20)_____ calories a day. The nursing mother should

drink about (21)_____ cups of liquids each day to prevent dehydration. Breastfeeding is inadvisable if

milk is contaminated with (22)_____, drugs, or environmental pollutants.

One of the most important nutrients for infants, as for everyone, is (23)_____ and

(24)_____ milk excels as a source of nutrients for the young infant. With the exception of

(25)_____, breast milk provides all the nutrients a healthy infant needs for the first (26)_____ months of life.

Solid food additions to a baby's diet should begin sometime between (27)_____ months and should be governed by the baby's nutrient needs, physical (28)_____ to handle different forms of foods, and the need to detect and control allergic reactions. The early feeding of an infant should foster preferences that will support normal (29)_____ and health throughout life and, most importantly, support normal (30)_____ as the child grows.

Chapter Glossary

Crossword Puzzle:

Across:

4. abnormal glucose tolerance appearing during pregnancy
5. the chief protein in human breast milk
8. behavioral, cognitive, or central nervous system abnormalities associated with prenatal alcohol exposure
9. an uncommon and always fatal neural tube defect in which the brain fails to form
11. a group of nervous system abnormalities caused by interruption of the normal early development of the neural tube
12. iron-deficiency anemia caused by drinking so much milk that iron-rich foods are displaced from the diet
14. a milklike secretion from the breasts during the first day or so after delivery before milk appears; rich in protective factors

Down:

1. the cluster of symptoms, including brain damage, growth retardation, mental retardation, and facial abnormalities, seen in an infant or child whose mother consumed alcohol during her pregnancy
2. a potentially dangerous condition during pregnancy characterized by edema, hypertension, and protein in the urine
3. one of the most common types of neural tube defects in which gaps in the bones of the spine often allow the spinal cord to bulge and protrude through the gaps, resulting in a number of motor and other impairments
6. production and secretion of breast milk for the purpose of nourishing an infant
7. surgical childbirth, in which the infant is taken through an incision in the woman's abdomen
10. a factor in breast milk that binds iron and keeps it from supporting the growth of the infant's intestinal bacteria
13. accumulation of fluid in the tissues

Word Bank:

alpha-lactalbumin
anencephaly
ARND
cesarean section
colostrum

edema
fetal alcohol syndrome
gestational diabetes
lactation
lactoferrin

milk anemia
neural tube defects
preeclampsia
spina bifida

Matching Exercise:

_____ 1. critical period

_____ 2. gestation

_____ 3. amniotic sac

_____ 4. fetus

_____ 5. ovum

_____ 6. embryo

_____ 7. zygote

_____ 8. uterus

_____ 9. placenta

_____ 10. certified lactation consultant

_____ 11. implantation

_____ 12. low birthweight

_____ 13. Special Supplemental Food Program for Women, Infants, and Children (WIC)

_____ 14. environmental tobacco smoke (ETS)

_____ 15. alcohol-related birth defects (ARBD)

_____ 16. trimester

_____ 17. listeriosis

a. the stage of human gestation from the third to eighth week after conception

b. a health care provider with specialized training and certification in breast and infant anatomy and physiology who teaches the mechanics of breastfeeding to new mothers

c. the period of about 40 weeks (three trimesters) from conception to birth; the term of a pregnancy

d. the stage of human gestation from eight weeks after conception until birth of an infant

e. a birthweight of less than 5 ½ pounds (2,500 grams); used as a predictor of probable health problems in the newborn and as a probable indicator of poor nutrition status of the mother before and/or during pregnancy

f. the womb, the muscular organ within which the infant develops before birth

g. a finite period during development in which certain events may occur that will have irreversible effects on later developmental stages

h. the stage of development, during the first two weeks after conception, in which the fertilized egg embeds itself in the wall of the uterus and begins to develop

i. the term that describes the product of the union of ovum and sperm during the first two weeks after fertilization

j. the organ of pregnancy in which maternal and fetal blood circulate in close proximity and exchange nutrients and oxygen and wastes

k. the egg, produced by the mother, that unites with a sperm from the father to produce a new individual

l. the "bag of waters" in the uterus in which the fetus floats

m. malformations in the skeletal and organ systems associated with prenatal alcohol exposure

n. the combination of exhaled smoke (mainstream smoke) and smoke from lighted cigarettes, pipes, or cigars (sidestream smoke) that enters the air and may be inhaled by other people

o. a USDA program offering low-income pregnant women or those with infants coupons redeemable for specific foods to supply the nutrients deemed most needed for growth and development

p. a serious foodborne infection caused by a bacterium found in soil and water that can cause severe brain infection or death in a fetus or newborn

q. a period representing gestation; about 13-14 weeks

Exercises

Answer these chapter study questions:

1. Why is an appropriate prepregnancy weight important and how does it relate to pregnancy outcome?

2. Do all women need to add protein-rich foods to their diets in order to obtain the additional daily protein recommended during pregnancy? Why or why not?

3. How does a woman's body conserve iron during pregnancy? Does this mean that an iron supplement should not be taken?

4. What effect does nutritional deprivation of the mother have on breastfeeding?

5. Discuss ways to prevent infant obesity and to encourage eating habits that will support continued normal weight as the child grows.

6. Why are pregnant women advised against eating large ocean fish, such as king mackerel and sharks?

Complete these short answer questions:

1. Maternal factors associated with low-birthweight infants include:

   a.   d.

   b.   e.

   c.

2. The effects of malnutrition during critical periods of pregnancy are seen in:

   a.

   b.

   c.

3. The pregnant woman's need for these two vitamins increases due to their roles in cell reproduction:

   a.   b.

4. These minerals, involved in the normal development of teeth and bones, are in great demand during pregnancy:

   a.   c.

   b.

5. Prenatal supplements typically provide more of these nutrients than regular supplements:

   a.   c.

   b.

6. Components of weight gained by the pregnant woman include these lean tissues:

   a.   d.

   b.   e.

   c.

7. The nausea of morning sickness can be alleviated by starting the day with:

   a.

   b.

8. Preeclampsia is characterized by:

   a.   c.

   b.

9. The addition of foods to a baby's diet should be governed by these three conditions:

   a.

   b.

   c.

10. Normal dental development in a baby should be promoted by:

   a.

   b.

   c.

   d.

Solve these problems:

1. How many calories and grams of carbohydrate and protein should be provided in a diet for a 30-year-old pregnant woman who weighs 130 pounds and is at her appropriate weight for height? She is in her second trimester of pregnancy. Her nonpregnant caloric needs are 2000 calories per day.

For questions 2-6, match the nutrients needed in extra amounts during pregnancy, listed on the right, with some of their best food sources, listed on the left.

_____	2. dried fruits	a. folate
_____	3. dark, green leafy vegetables	b. calcium
		c. iron
_____	4. eggs	d. zinc
_____	5. milk	e. vitamin B_{12}
_____	6. shellfish	

For questions 7-12, identify the correct sequence for introducing the following foods into an infant's diet, by placing an "a" by the first food which should be introduced, a "b" by the second food to be introduced, etc.

_____ 7. pureed vegetables

_____ 8. pieces of soft, cooked vegetables from the table

_____ 9. yogurt

_____ 10. iron-fortified rice cereal

_____ 11. mashed fruits

_____ 12. finely-cut meats

Answer these controversy questions:

1. Describe the typical characteristics of children most likely to be obese.

2. What are characteristics of children at the highest risk of developing heart disease?

3. What are characteristics of physically active children versus sedentary children?

4. Describe recommended dietary guidelines for children older than two that benefit blood lipids without compromising the child's growth and development.

5. What are recommendations for controlling hypertension in children?

Study Aids

1. Complete the following table by filling in the blanks.

First Foods for the Infant	
Age (months)	**Addition**
0-4	Breast milk or (a)_____
4-6	Iron fortified (b)_____ mixed with breast milk, formula, or water
6-8	Begin plain baby food (c)_____, mashed (d)_____ and fruits
8-10	Gradually begin yogurt, finely cut (e)_____, fish, casseroles, cheese, (f)_____, and legumes

2. Use the following tables in your textbook as study aids: study Table 13-2 on page 493 to identify daily food choices for pregnant and lactating women; review Table 13-4 on page 497 for the recommended weight gains for pregnancy; and study Table 13-8 on page 514 to compare and contrast the composition of human milk versus infant formula.

Sample Test Items

Comprehension Level Items:

1. Which of the following is the major factor in low birthweight?

   a. heredity   c. drug use
   b. smoking   d. poor nutrition

2. The muscular organ within which the infant develops before birth is called the

   a. placenta.   c. ovum.
   b. uterus.   d. zygote.

3. The stage of human gestation from the third to eighth week after conception is called a(an):

   a. ovum.   c. embryo.
   b. zygote.   d. fetus.

4. The recommended protein intake during pregnancy is _____ grams per day higher than for nonpregnant women.

   a. 25
   b. 50
   c. 60
   d. 100

5. Folic acid, the synthetic form of folate, is better absorbed than the naturally occurring folate in foods.

   a. true
   b. false

6. A daily iron supplement containing _____ milligrams is recommended during the second and third trimesters of pregnancy.

   a. 10
   b. 20
   c. 30
   d. 40

7. A sudden large weight gain during pregnancy is a signal that may indicate:

   a. that the woman is consuming too many calories.
   b. that the baby will be too large.
   c. the onset of preeclampsia.
   d. a and b
   e. b and c

8. Which of the following practices should be avoided during pregnancy?

   a. exercise
   b. dieting
   c. moderate consumption of caffeine
   d. a and b
   e. b and c

9. A strategy which has been found effective in alleviating the nausea of morning sickness is to start the day with:

   a. a few sips of a carbonated beverage.
   b. a food high in fat.
   c. a salty snack food.
   d. a and b
   e. a and c

10. Breast milk volume depends on:

    a. the size of the mother's breasts.
    b. how much fluid the mother consumes.
    c. how much milk the baby demands.
    d. a and b
    e. b and c

11. Characteristics of breast milk include all of the following **except**

    a. its carbohydrate is in the form of lactose.
    b. it is high in sodium.
    c. its iron is highly absorbable.
    d. its zinc is absorbed better than zinc from cow's milk.

12. A baby less than one year of age who has been on breast milk should be weaned onto:

    a. whole milk.   c. low-fat milk.

    b. skim milk.   d. infant formula.

13. At 6 months of age, an exclusively breastfed baby needs additional:

    a. vitamin C.   c. iron.

    b. vitamin D.   d. vitamin A.

14. Food aversions and cravings that arise during pregnancy are usually due to:

    a. changes in taste.   d. a and b

    b. physiological needs.   e. a and c

    c. changes in smell sensitivities.

15. Birth defects have been reliably observed in the children of some women who drank _____ ounces of alcohol daily during pregnancy.

    a. 2   c. 4

    b. 3   d. 5

16. Producing 25 ounces of milk a day costs a woman almost _____ calories per day.

    a. 300   c. 650

    b. 500   d. 1000

17. The birthweight of an infant triples by the age of:

    a. 4 months.   c. 9 months.

    b. 6 months.   d. 1 year.

18. Reduced-fat milk is not recommended for infants and children before the age of:

    a. 6 months.   c. 3 years.

    b. 2 years.   d. 4 years.

19. Supplemental water should be offered regularly to both breastfed and bottle-fed infants.

    a. true   b. false

20. WIC participation during pregnancy can effectively reduce infant mortality, low birthweight, and maternal and newborn medical costs.

    a. true   b. false

21. All of the following are appropriate physical activities for the pregnant woman **except**:

    a. water aerobics.   c. climbing stairs.

    b. light strength training.   d. volleyball.

22. The greatest risk for pregnant teenagers is:

    a. having low-birthweight infants.
    b. the death of the infant within the first year.
    c. experiencing a miscarriage.
    d. having a premature baby.

23. Breastfeeding is inadvisable if the mother:

    a. is drinking alcohol.   d. a and b
    b. has a cold.   e. a and c
    c. is on drugs.

24. Breastfeeding protects against common illnesses of infancy such as middle ear infection and respiratory illness.

    a. true   b. false

Application Level Items:

25. Lola experienced malnutrition late in her pregnancy. Which of the following organs of the infant will most likely be affected?

    a. heart   d. a and c
    b. lungs   e. b and c
    c. brain

26. Cindy is a 25-year-old pregnant female who recently experienced a sudden weight gain. In addition, she has blurred vision, is dizzy, and has frequent headaches. Cindy is most likely experiencing symptoms of:

    a. malnutrition.   c. preeclampsia.
    b. gestational diabetes.   d. a normal pregnancy.

27. A nonpregnant woman requires 2200 calories to maintain her desirable body weight. How many calories would she need if she were in her third trimester of pregnancy?

    a. 2100   c. 2500
    b. 2300   d. 2650

28. Joan is a strict vegetarian who is now pregnant. Which of the following would you recommend as appropriate sources of protein?

    a. legumes   d. a and b
    b. tofu   e. b and c
    c. protein supplements

29. Jane weighs 160 pounds (prepregnancy) and is considered to be obese. Approximately how much should Jane weigh at the end of her pregnancy?

    a. 160   c. 180
    b. 175   d. 190

30. Which of the following indicates an infant who is ready for solid foods?

    a. Jeff, who is 4 months old
    b. Claudia, who can sit with support and control head movements
    c. Ben, who swallows using the back of his tongue
    d. Sarah, who turns her head toward any object that brushes her cheek

31. Which of the following would you do to encourage a baby to sleep through the night?

    a. give the baby warm milk right before bedtime
    b. put the baby to bed with a bottle of milk
    c. stuff the baby with solid foods right before bedtime
    d. do nothing, since babies sleep through the night whenever they are ready

32. Which of the following foods would you introduce first to a baby to prevent food allergies?

    a. egg whites   c. rice cereal
    b. citrus fruits   d. cow's milk

Answers

Summing Up

1. nourish
2. weight
3. birthweight
4. health
5. placenta
6. sustenance
7. critical
8. heart
9. digestive
10. 40
11. birth
12. 340
13. 25
14. carbohydrate
15. folate
16. phosphorus
17. iron
18. three
19. pound
20. 500
21. 13
22. alcohol
23. water
24. breast
25. vitamin D
26. six
27. 4-6
28. readiness
29. development
30. weight

Chapter Glossary

Crossword Puzzle:
1. fetal alcohol syndrome
2. preeclampsia
3. spina bifida
4. gestational diabetes
5. alpha-lactalbumin
6. lactation
7. cesarean section
8. ARND
9. anencephaly
10. lactoferrin
11. neural tube defects
12. milk anemia
13. edema
14. colostrum

Matching Exercise:
1. g
2. c
3. l
4. d
5. k
6. a
7. i
8. f
9. j
10. b
11. h
12. e
13. o
14. n
15. m
16. q
17. p

Exercises

Chapter Study Questions:

1. A strong correlation exists between prepregnancy weight, weight gain during pregnancy, and infant birthweight. Infant birthweight is the most potent single indication of the infant's future health status. Low-birthweight babies have a greater chance of dying early in life and of having illnesses.

2. No; many women in the United States exceed the recommended protein intake for pregnancy even when they are not pregnant and excessive protein intake may have adverse effects.

3. Menstruation ceases and the absorption of iron increases up to threefold. However, because iron stores dwindle, blood is lost during birth, and few women enter pregnancy with adequate stores to meet pregnancy demands, a supplement is recommended during the second and third trimesters.

4. In general, the effect of nutritional deprivation is to reduce the quantity, not the quality, of milk. However, the levels of some vitamins in human milk can be affected by excessive or deficient intakes of the mother. Nutrients in breast milk most likely to be affected by prolonged inadequate intake of the mother are vitamins B_6, B_{12}, A and D.

5. A variety of nutritious foods should be introduced in an inviting way; the baby should not be forced to finish a bottle or jar of baby food; concentrated sweets and empty-calorie foods should be limited; babies should not be taught to seek food as a reward and they should not be comforted or punished with food; physical activity should be encouraged.

6. Because they contain high concentrations of mercury which is an environmental contaminant that can harm the developing brain and nervous system.

Short Answer Questions:

1. (a) poor nutrition; (b) heredity; (c) disease conditions; (d) smoking; (e) drug use, including alcohol

2. (a) defects of the nervous system of the embryo; (b) poor dental health of children whose mothers were malnourished during pregnancy; (c) adult's and adolescent's vulnerability to infections and possibly higher risks of diabetes, hypertension, stroke and heart disease

3. (a) folate; (b) vitamin B_{12}

4. (a) calcium; (b) phosphorus; (c) magnesium

5. (a) folate; (b) iron; (c) calcium

6. (a) placenta; (b) uterus; (c) blood; (d) milk-producing glands; (e) fetus itself

7. (a) a few sips of a carbonated drink; (b) a few nibbles of soda crackers or other salty snack foods

8. (a) hypertension; (b) edema; (c) protein in the urine

9. (a) the baby's nutrient needs; (b) the baby's physical readiness to handle different forms of foods; (c) need to detect and control allergic reactions

10. (a) supplying nutritious foods; (b) avoiding sweets; (c) discouraging the association of food with reward or comfort; (d) discouraging the use of a bottle as a pacifier

Problem-Solving:

1. (a) caloric needs: 2000 calories + 340 calories = 2340 calories; (b) recommended carbohydrate intake = at least 175 grams (263-380 grams based on the AMDR); (c) recommended protein intake = .8 g/kg body weight + 25 additional grams of protein = .8 × 59 = 47 + 25 = 72 grams

2. c
3. a
4. e
5. b

6. d
7. b
8. e
9. d

10. a
11. c
12. f

Controversy Questions:

1. They are female, have a family history of type 2 diabetes, are of non-European descent, have mothers who had diabetes while pregnant, have metabolic syndrome, have a low family income, are sedentary, and have parents who are obese.

2. Children with the highest risk are obese, are sedentary, and may have diabetes, high blood pressure, and high LDL blood cholesterol.

3. Physically active children have a better lipid profile and lower blood pressure than sedentary children. They also have reduced waist circumference, improved muscle strength, and better condition of the heart and arteries.

4. A diet limited in fat, especially saturated fat, *trans* fat and cholesterol, rich in nutrients, and age-appropriate in calories is recommended for heart health.

5. Participation in regular aerobic activity and losing weight or maintaining weight as children grow taller. Also, restricting sodium intake and watching intake of caffeinated beverages may help decrease blood pressure.

Study Aids

1. (a) infant formula; (b) cereal; (c) meats; (d) vegetables; (e) meats; (f) eggs

Sample Test Items

1. d (p. 488)
2. b (p. 489)
3. c (p. 488)
4. a (p. 493)
5. a (pp. 494-495)
6. c (p. 496)
7. c (p. 506)
8. b (pp. 502-503)

9. e (p. 500)
10. c (p. 507)
11. b (p. 512)
12. d (p. 514)
13. c (p. 512)
14. e (p. 499)
15. a (p. 505)
16. b (p. 507)

17. d (p. 509)
18. b (p. 515)
19. b (p. 510)
20. a (p. 497)
21. d (pp. 498-499)
22. b (p. 499)
23. e (p. 508)
24. a (p. 513)

25. b (p. 490)
26. c (p. 506)
27. d (p. 491)
28. d (p. 493)
29. b (p. 497)
30. b (p. 516)
31. d (p. 517)
32. c (p. 517)

people, but (27)_____ deficiencies are common in older people and may lead to decreased appetite and a diminished sense of (28)_____. Preventing (29)_____ is an important concern for the older person suffering with Alzheimer's disease, due to skipped meals and poor food choices. Two factors that make older people vulnerable to malnutrition include use of multiple (30)_____ and abuse of alcohol.

Exercises

Answer these chapter study questions:

1. Identify four ways in which television affects children's nutritional health adversely.

2. Compare the characteristics of children who eat no breakfast with those of their well-fed peers.

3. What recommendations would you make to an individual who experiences PMS?

4. Why should older adults attend to fat intakes?

5. Why do energy needs decrease with age?

6. How would you prevent a child from choking?

Complete these short answer questions:

1. School lunches must include specified servings of:

   a.   d.

   b.   e.

   c.

2. Children absorb more lead if they are lacking in these nutrients:

   a.   d.

   b.   e.

   c.

3. The life threatening food allergy reaction of anaphylac tic shock is most often caused by these foods:

   a.   e.

   b.   f.

   c.   g.

   d.   h.

4. Normal, everyday ca uses of cranky or rambunctious children include:

   a.

   b.

   c.

   d.

   e.

   f.

5. Hormones that regulate the menstrual cycle also alter:

    a.   d.

    b.   e.

    c.   f.

6. Children prefer vegetables that meet these descriptions:

    a.

    b.

    c.

    d.

7. Older people suffer more deficiencies of these two vitamins than younger people:

    a.   b.

8. Older people face a greater risk of vitamin D deficiency than younger people do because:

    a.

    b.

    c.

9. Six factors that affect physiological age include:

    a.

    b.

    c.

    d.

    e.

    f.

10. The defining symptom of Alzheimer's disease is impairment of memory and reasoning powers, but it may also be accompanied by:

    a.   f.

    b.   g.

    c.   h.

    d.   i.

    e.

Solve these problems:

1. Below is a sample day's menu for a child who needs about 1,800 calories per day. Compare the menu with MyPyramid for Kids and identify foods and potential nutrients which would be missing from the child's diet.

    Breakfast: Cheese toast with 1 slice of whole-wheat bread and 1 ounce of cheddar cheese; ¾ cup orange juice; ½ cup whole milk

    Lunch: Peanut butter and jelly sandwich with 4 tbsp. peanut butter and 2 tbsp. jelly on 2 slices of whole-wheat bread; 1 apple; 1 cup whole milk

    Dinner: 2 ounces of steak; ½ cup mashed potato with 1 tbsp. butter; 1 dinner roll; 1 cup strawberry yogurt

    Snack: 3 chocolate chip cookies

2. Identify whether the following foods have high or low caries potential by placing an X in the appropriate space.

		Low Caries Potential	High Caries Potential
a.	plain yogurt	_____	_____
b.	sugared cereals	_____	_____
c.	bagels	_____	_____
d.	dried fruit	_____	_____
e.	popcorn	_____	_____
f.	glazed carrots	_____	_____
g.	sugarless gum	_____	_____
h.	legumes	_____	_____
i.	fresh fruit	_____	_____
j.	chocolate milk	_____	_____

3. For each nutrient listed below, indicate whether nutritional needs for older adults are higher, lower, or the same as that for younger adults.

		Higher	Lower	Same
a.	energy	_____	_____	_____
b.	fiber	_____	_____	_____
c.	vitamin A	_____	_____	_____
d.	vitamin D	_____	_____	_____
e.	protein	_____	_____	_____

8. What is the main source of lead in most children's lives?

    a. old paint   c. water pipes
    b. food cans   d. gasoline

9. Which of the following foods have been shown not to worsen acne?

    a. chocolate   d. a and b
    b. fatty acids   e. a and c
    c. sugar

10. Which of the following would you recommend to a woman who experiences PMS?

    a. a diuretic   c. a caffeine-free lifestyle
    b. vitamin E supplement   d. a fiber-restricted diet

11. Which of the following protein sources would you recommend to an older adult with constipation?

    a. beef   c. legumes
    b. eggs   d. milk

12. In the United States, the life expectancy for black men is almost:

    a. 70 years.   c. 80 years.
    b. 75 years.   d. 90 years.

13. A diet low in _____ may improve some symptoms of rheumatoid arthritis.

    a. protein   c. sugar
    b. sodium   d. saturated fat

14. Energy needs often decrease with advancing age.

    a. true   b. false

15. As a person grows older:

    a. dehydration becomes a major risk.   d. a and b
    b. fiber takes on extra importance.   e. b and c
    c. protein needs increase.

16. Which of the following statements is true concerning the dietary habits of older adults?

    a. They have cut down on saturated fats.
    b. They are eating fewer vegetables.
    c. They are consuming more milk.
    d. They are eating fewer whole-grain breads.

17. There is an apparent increase in the absorption of vitamin _____ with aging.

    a. C   c. A
    b. D   d. E

18. Which of the following is provided by the Senior Nutrition Program?

    a.   nutritious meals   d.   a and b
    b.   transportation   e.   b and c
    c.   assistance with housekeeping

19. Which of the following puts older adults at greatest risk of developing cataracts according to the latest scientific studies?

    a.   excess food energy intake
    b.   diets low in fruits and green vegetables
    c.   excess intakes of milk and sugar
    d.   deficiencies of riboflavin and zinc

20. Students who regularly eat school lunches have higher intakes of many nutrients and of fiber than those who don't.

    a.   true   b.   false

21. To determine the appropriate fiber intake for children you would add the number _____ to their age.

    a.   2   c.   4
    b.   3   d.   5

22. Which of the following are ways to prevent iron deficiency in children?

    a.   Provide 10 milligrams of iron per day from foods.
    b.   Encourage the consumption of four cups of milk per day.
    c.   Serve iron-rich foods at meals and snacks.
    d.   a and b
    e.   a and c

23. Research has shown a positive correlation between childhood obesity and fruit juice consumption.

    a.   true   b.   false

24. Teenagers often choose _____ as their primary beverage.

    a.   water   c.   soft drinks
    b.   milk   d.   fruit juice

Application Level Items:

25. Which of the following would be the best snack choice for an active, normal-weight child?

    a.   flavored gelatin   c.   cola
    b.   candy   d.   oatmeal cookies

26. Which of the following vegetables would be the most acceptable to a child?

    a.   slightly overcooked green beans   c.   broccoli
    b.   mixed vegetables   d.   bright green peas

27. Betsy is a two year old who has been on a food jag for five days. What is the best way to respond to Betsy's food jag?

    a. no response, since attention is a valuable reward
    b. serve tiny portions of many foods
    c. serve favored food items
    d. distract the child with friends at meals

28. Which foods should be used infrequently in order to prevent dental caries?

    a. cheese, apples, and hard rolls
    b. toast, chicken, and green beans
    c. chocolate milk, raisins, and candied sweet potatoes
    d. plain yogurt, eggs, and pizza

29. Which of the following snack foods would be most appropriate for a teenager trying to consume enough iron?

    a. meat sandwich, low-fat bran muffin, and orange juice
    b. carrot sticks, tomato juice, and dried apricots
    c. hard-boiled egg, whole-wheat bread, and celery sticks
    d. Swiss cheese, crackers, and cantaloupe

30. Bessie is an 85-year-old female who resides in a nursing home. She has a good appetite and consumes three cups of milk a day. However, Bessie is at risk for developing vitamin D deficiency because:

    a. she does not have adequate exposure to sunlight.
    b. she does not consume an adequate amount of milk.
    c. as she ages, her vitamin D synthesis declines.
    d. a and b
    e. a and c

Answers

Chapter Glossary

Matching Exercise:

1. f	5. d	9. a	13. e	17. p
2. t	6. o	10. m	14. n	18. s
3. b	7. g	11. h	15. c	19. r
4. l	8. q	12. k	16. i	20. j

Summing Up

1. gender
2. first
3. 800
4. 800
5. 10
6. caffeine
7. hunger
8. television
9. adolescence
10. fat

11. snacks
12. menstruation
13. appetite
14. behavior
15. iron
16. calcium
17. stress
18. 81
19. energy
20. carbohydrate

21. fats
22. dehydration
23. fiber
24. A
25. D
26. vegetable
27. zinc
28. taste
29. weight loss
30. medications

Exercises

Chapter Study Questions:

1. Television viewing requires no energy above the resting level of expenditure. It contributes to physical inactivity by consuming time that could be spent in more vigorous activities. Watching television also correlates with between-meal snacking and with eating the calorically dense, fatty and sugary foods most heavily advertised on children's programs. Finally, children who watch more television or watch it during meals are least likely to eat fruits and vegetables and more likely to be obese.

2. Children who eat no breakfast are more likely to be overweight and to perform poorly in tasks of concentration, their attention spans are shorter, they achieve lower test scores, and they are tardy or absent more often than their well-fed peers. They are also more likely to snack on sweet and salty foods.

3. The individual should try to examine her total lifestyle, including diet and stress, get adequate sleep, engage in physical activity, and try to moderate her intake of caffeine.

4. Foods low in saturated and *trans* fats are often richest in vitamins, minerals and phytochemicals and may help retard the development of atherosclerosis, obesity, and other diseases. A diet low in the fats of meats and dairy products may also improve some symptoms of arthritis. In addition, consuming enough of the essential fatty acids supports good health throughout life.

5. The number of active cells in each organ decreases, which brings about a reduction in the body's metabolic rate. In addition, older people are not usually as physically active as younger people and their lean tissue diminishes.

6. Have the child sit while eating and eliminate round foods, such as grapes and hard candies, or other dangerous foods, such as popcorn, tough meat, and peanut butter eaten by the spoonful.

Short Answer Questions:

1. (a) milk; (b) protein-rich foods; (c) vegetables; (d) fruits; (e) breads or other grains

2. (a) calcium; (b) zinc; (c) vitamin C; (d) vitamin D; (e) iron

3. (a) milk; (b) eggs; (c) peanuts; (d) tree nuts; (e) wheat; (f) soybeans; (g) fish; (h) shellfish

4. (a) desire for attention; (b) lack of sleep; (c) overstimulation; (d) lack of exercise; (e) chronic hunger; (f) too much caffeine from colas or chocolate

5. (a) the metabolic rate; (b) glucose tolerance; (c) appetite; (d) food intake; (e) mood; (f) behavior

6. (a) mild flavored; (b) slightly undercooked and crunchy; (c) brightly colored; (d) easy to eat

7. (a) vitamin D; (b) vitamin B_{12}

8. (a) many drink little or no vitamin D-fortified milk; (b) many get little or no exposure to sunlight; (c) as people age, vitamin D synthesis declines

9. (a) abstinence from, or moderation in, alcohol use; (b) regular nutritious meals; (c) weight control; (d) adequate sleep; (e) abstinence from smoking; (f) regular physical activity

10. (a) anxiety; (b) loss of ability to communicate; (c) loss of physical capabilities; (d) delusions; (e) depression; (f) inappropriate behavior; (g) irritability; (h) sleep disturbance; (i) eventual loss of life

Problem-Solving:

1. About 3 cups were consumed from the milk group, which meets the recommended amount. Six ounce-equivalents of meat and beans were consumed, 1 ounce more than recommended. The grains group was not adequately represented, with a total of four ounces versus the recommended six. About 1 ¼ cup-equivalents of fruits were consumed, which was only ¼ cup short of meeting the goal, and only ½ cup of vegetables was consumed, versus the 2 ½ cups recommended. As a consequence, vitamins A and C, folate, carbohydrate, and fiber intake could be low for the day.

2.

		Low Caries Potential	High Caries Potential
a.	plain yogurt	X	
b.	sugared cereals		X
c.	bagels	X	
d.	dried fruit		X
e.	popcorn	X	
f.	glazed carrots		X
g.	sugarless gum	X	
h.	legumes	X	
i.	fresh fruit	X	
j.	chocolate milk		X

3.

		Higher	Lower	Same
a.	energy		X	
b.	fiber			X
c.	vitamin A		X	
d.	vitamin D	X		
e.	protein			X

Controversy Questions:

1. Those who take drugs for long times; those who take 2 or more drugs at the same time; and those who are poorly nourished to begin with or not eating well.

2. Laxatives can carry nutrients through the intestines so rapidly that many vitamins have no time to be absorbed. Also, mineral oil can rob a person of fat-soluble vitamins, and calcium can also be excreted along with the oil, leading to adult bone loss.

3. Aspirin speeds up blood loss from the stomach by as much as ten times, leading to iron-deficiency anemia.

4. Because oral contraceptives alter blood lipids by raising total cholesterol and triglyceride concentrations and by lowering HDL. Some women also experience mild hypertension.

5. Moderation is defined as the equivalent of 2 small cups of coffee per day.

Sample Test Items

1. b (p. 532)	9. e (p. 547)	17. c (p. 553)	25. d (p. 532)
2. d (p. 541)	10. c (p. 548)	18. d (p. 559)	26. d (p. 534)
3. c (p. 538)	11. c (p. 552)	19. b (p. 554)	27. a (p. 534)
4. a (p. 542)	12. a (p. 550)	20. a (p. 542)	28. c (p. 541)
5. c (p. 545)	13. d (p. 553)	21. d (p. 531)	29. a (p. 549)
6. a (p. 548)	14. a (p. 550)	22. e (p. 532)	30. e (p. 553)
7. e (p. 534)	15. d (pp. 552-554)	23. b (p. 533)	
8. a (p. 537)	16. a (p. 559)	24. c (p. 545)	

Chapter Glossary

Matching Exercise:

_____ 1. food poverty

_____ 2. famine

_____ 3. emergency kitchens

_____ 4. world food supply

_____ 5. food recovery

_____ 6. hunger

_____ 7. sustainable

_____ 8. oral rehydration therapy

_____ 9. carrying capacity

_____ 10. biofuels

_____ 11. food banks

_____ 12. food pantries

a. a consequence of food insecurity that, because of prolonged involuntary lack of food, results in discomfort, illness, weakness, or pain beyond a mild uneasy sensation

b. oral fluid replacement for children with severe diarrhea caused by infectious disease

c. widespread and extreme scarcity of food that causes starvation and death in a large portion of the population in an area

d. community food collection programs that provide groceries to be prepared and eaten at home

e. fuels made mostly of materials derived from recently harvested living organisms

f. the quantity of food, including stores from previous harvests, available to the world's people at a given time

g. able to continue indefinitely; the term refers to the use of resources at such a rate that the earth can keep on replacing them

h. hunger occurring when enough food exists in an area but some of the people cannot obtain it because they lack money, are being deprived for political reasons, live in a country at war, or suffer from other problems such as lack of transportation

i. the total number of living organisms that a given environment can support without deteriorating in quality

j. programs that provide prepared meals to be eaten on-site

k. collecting wholesome surplus food for distribution to low-income people who are hungry

l. facilities that collect and distribute food donations to authorized organizations feeding the hungry

Exercises

Answer these chapter study questions:

1. Explain why undernutrition and obesity often occur together within the same family or person.

2. What factors are contributing to loss of food-producing land around the world?

3. Why are fast cooking methods recommended as one step consumers should take to minimize negative impacts on the environment? What are some examples?

4. Describe what is meant by the statement that hunger, poverty, and environmental degradation all interact with each other.

5. Why do people living in poverty bear more children?

6. Why are women being targeted as direct recipients of food relief?

Complete these short answer questions:

1. Some of the forces threatening world food production and distribution in the next decades include:

   a.

   b.

   c.

   d.

   e.

   f.

   g.

2. The primary cause of hunger in developed nations is _____.

244

5. What is the U.S. Conservation Reserve Program?

Sample Test Items

Comprehension Level Items:

1. Which of the following forces are expected threaten world food production and distribution in the next decades?

    a. The world's supply of fresh water is dwindling.
    b. Global warming is decreasing.
    c. The ozone layer is growing thicker.
    d. The amount of food-producing land is increasing.

2. Low food security is characterized by:

    a. reduced dietary quality, variety or desirability.
    b. significant reduction in total food intake.
    c. a diet centered on inexpensive, low-nutrient dense foods.
    d. a and b
    e. a and c

3. The term *sustainable* means able to continue indefinitely.

    a. true   b. false

4. What is the major cause of hunger in the United States?

    a. famine   c. food shortage
    b. food poverty   d. food insecurity

5. Which of the following is the leading cause of blindness in the world's young children?

    a. vitamin A deficiency   c. vitamin C deficiency
    b. iodine deficiency   d. protein deficiency

6. If you are a meat eater, which of the following recommendations would you follow in order to be environmentally conscious?

    a. Select canned corn beef.   d. a and b
    b. Choose chicken from local farms.   e. b and c
    c. Select small fish.

7. Overpopulation may be the most serious threat that humankind faces today.

    a. true   b. false

8. Which of these types of bags is the best choice from an environmental perspective?

   a. paper
   b. plastic
   c. cloth

   d. a and b
   e. b and c

9. Locally grown foods require less transportation, packaging, and refrigeration than shipped foods.

   a. true

   b. false

10. What is the single largest source of air pollution?

    a. motor vehicles
    b. agriculture

    c. manufacturing industries
    d. paper mills

11. Which of the following cooking aids is the best choice from an environmental perspective?

    a. paper towels
    b. plastic wrap

    c. dishcloths
    d. aluminum foil

12. Products bearing the U.S. government's Energy Star logo rank highest for energy efficiency.

    a. true

    b. false

13. In the developing world, the primary form of hunger is:

    a. food insecurity.
    b. food shortage.

    c. famine.
    d. food poverty.

14. In the United States, one of every _____ Americans receives food assistance of some kind.

    a. 2
    b. 4

    c. 6
    d. 10

15. Hunger is not always easy to recognize.

    a. true

    b. false

16. To help solve the world's environmental, poverty, and hunger problems, the poor nations should:

    a. adopt sustainable practices that slow and reverse destruction of their forests, waterways and soil.
    b. educate and assist the poor.
    c. stem their wasteful and polluting use of resources and energy.
    d. a and b
    e. b and c

17. The centerpiece of U.S. food programs for low-income people is the:

    a. WIC Program.
    b. National School Lunch Program.

    c. Emergency Food Assistance Program.
    d. Food Stamp Program.

18. Recipients can use food stamps to purchase:

    a. alcohol.
    b. cleaning items.

    c. seeds.
    d. tobacco.

19. Government programs are helpful in relieving poverty and hunger.

    a. true   b. false

20. Which of the following is a prerequisite to curbing population growth?

    a. relieving poverty
    b. inventing effective birth control measures
    c. alleviating hunger
    d. a and b
    e. a and c

21. Most children who die of malnutrition do so because:

    a. they starve to death from lack of adequate food intake.
    b. their mothers keep the food for themselves and their fathers.
    c. they lack adequate high-quality protein from animal sources.
    d. their health has been compromised by dehydration from infections that cause diarrhea.

22. The numbers of people affected by famine are relatively small compared with those suffering from less severe but chronic hunger.

    a. true   b. false

23. What nutrients are lacking in the diets of about half of the earth's population?

    a. vitamin A, vitamin C, and iron   c. zinc, protein, and vitamin C
    b. iodine, iron, and vitamin A   d. iodine, zinc, and vitamin D

24. About half of the deaths among the world's children are attributable to malnutrition.

    a. true   b. false

25. Consumption of these fish helps to protect over-fished species:

    a. Alaskan halibut.   d. a and b
    b. swordfish.   e. a and c
    c. white sea bass.

Application Level Items:

26. Which of the following would you do to be an environmentally conscious food shopper?

    a. Buy bulk items with minimal packaging.
    b. Choose beef over fish.
    c. Eat canned chili.
    d. Choose fish which has been flown in from far away .

27. By purchasing fresh foods grown locally you are:

    a. selecting nutritionally superior foods.
    b. saving money.
    c. helping the local economy.
    d. minimizing negative impacts on the environment.

28. You want to cook a whole roast for dinner. Which of the following would be the safest and best way to save energy?

    a. Use the microwave.
    b. Use the oven.

    c. Use the stove top.
    d. Use the pressure cooker.

29. To solve the world's environmental, poverty, and hunger problems, rich nations must:

    a. gain control of their population growth.
    b. stem their wasteful and polluting uses of resources and energy.
    c. reverse the destruction of their environmental resources.
    d. educate their citizens.

30. You know a family whose diet supplies an abundance of calories primarily from white bread and crackers, cheap cuts of meat, and foods high in fat and sugar. They also consume very few fruits, vegetables and milk products. Which USDA food security term would you use to describe this family?

    a. marginal food security
    b. low food security

    c. very low food security
    d. high food security

Answers

Summing Up

1. 6
2. food insecurity
3. food poverty
4. obesity
5. developing
6. environmental
7. soil erosion
8. overpopulation
9. more
10. poverty
11. hunger
12. educate
13. sustainable
14. resources
15. economies
16. poverty
17. 2015
18. agricultural technology
19. environment
20. shopping
21. food chain
22. packaging
23. fast
24. appliances
25. Energy Star

Chapter Glossary

Matching Exercise:

1. h
2. c
3. j
4. f
5. k
6. a
7. g
8. b
9. i
10. e
11. l
12. d

Exercises

Chapter Study Questions:

1. The most affordable and available foods provide abundant calories, but few nutrients; examples are refined grains, sweets, inexpensive meat, oils and fast foods. In addition, people who have gone hungry in the past and whose future meals are uncertain may overeat when food or money becomes available.

2. Soil erosion is affecting agriculture in every nation, primarily due to deforestation of the world's rainforests. Land is also becoming saltier due to continuous irrigation which leaves deposits of salt in the soil and results in lower crop yields.

3. Fast cooking methods use less energy than conventional stovetop or oven cooking methods. Examples include stir-frying, pressure cooking, and microwaving.

4. Poor people often destroy the resources they need for survival, such as soil and forests. When they cut the trees for firewood or timber to sell, they lose the soil to erosion and, without these resources, they become poorer still. Thus, poverty causes environmental ruin and hunger grows from environmental ruin.

5. Poverty and hunger are correlated with lack of education which includes lack of knowledge on how to control family size. Also, families in poverty depend on children to farm the land, haul water, and care for adults in their old age. Many young children living in poverty also die from disease or other causes, in which case parents choose to have many children as a form of "insurance" that some will survive to adulthood.

6. Because in many nations women are the sole breadwinners in their household and because many women are also engaged in farming. Also, even starving women give what food is available to their children. Providing support to women also supports sustainable activities that improve the conditions and increases available food in the community.

Short Answer Questions:

1. (a) hunger, poverty, and population growth; (b) loss of food-producing land; (c) accelerating fossil fuel use; (d) atmosphere and climate changes, droughts, and floods; (e) ozone loss from the outer atmosphere; (f) water shortages; (g) ocean pollution

2. food poverty

3. (a) drought; (b) flood; (c) pests

4. (a) stir-frying; (b) pressure cooking; (c) microwaving

5. (a) stunted growth; (b) poor learning; (c) extreme weakness; (d) clinical signs of PEM; (e) increased susceptibility to disease; (f) loss of the ability to stand or walk; (g) premature death

6. (a) adopt sustainable development practices that slow and reverse the destruction of their forests, waterways, and soil; (b) educate and assist the poor

7. (a) low-interest loans; (b) basic equipment; (c) access to land ownership; (d) supports such as child care and education

8. (a) tobacco; (b) cleaning items; (c) alcohol; (d) nonfood items

9. stem their wasteful and polluting uses of resources

10. (a) field gleaning; (b) perishable food rescue or salvage; (c) prepared food rescue; (d) nonperishable food collection

Problem-Solving:

__X__   car pool to the grocery store
_____   eat high on the food chain
__X__   buy foods which are locally grown
_____   buy canned beef products
_____   select fish such as tuna, swordfish, and shark
__X__   buy items in bulk
_____   use plastic bags instead of paper ones
__X__   use the microwave for cooking

Controversy Questions:

1. Agriculture that is practiced on a massive scale by large corporations owning vast acreages and employing intensive technologies, fuel and chemical inputs.

2. By causing ozone depletion, water pollution, ocean pollution, and other ills and by making global warming likely.

3. Large masses of animal waste are produced and they leach into local soils and water supplies, polluting them. These animals also have to be fed and grain is grown for them on other land, which also requires fertilizers, herbicides, pesticides and irrigation.

4. Farmers using this system employ many techniques, such as crop rotation and natural predators, to control pests rather than relying on heavy use of pesticides alone.

5. It is a program that provides federal assistance to farmers who wish to improve their conservation of soil, water, and related natural resources on environmentally sensitive lands.

Sample Test Items

1. a (p. 580)	9. a (p. 586)	17. d (p. 576)	25. e (p. 581)
2. e (p. 572)	10. a (p. 586)	18. c (p. 576)	26. a (p. 586)
3. a (p. 584)	11. c (p. 586)	19. a (p. 576)	27. d (p. 586)
4. b (p. 573)	12. a (p. 586)	20. e (p. 582)	28. d (p. 586)
5. a (p. 578)	13. d (p. 576)	21. d (p. 579)	29. b (p. 584)
6. e (p. 586)	14. c (p. 574)	22. a (p. 578)	30. b (pp. 572-574)
7. a (p. 582)	15. a (p. 574)	23. b (p. 578)	
8. c (p. 586)	16. d (p. 584)	24. a (p. 578)	